2023
时装艺术国际先锋展
中国·濮院

介质律动

中国纺织工程学会　吕越　主编

中国纺织出版社有限公司

内 容 提 要

2023年的纺织时尚行业已经迎来了新的发展阶段，全球各地的文化交流复归频繁，多元文化的融合成为时尚艺术发展的重要动力。智能材料、可穿戴技术和虚拟现实等创新科技的应用，为时装艺术带来了前所未有的可能性。数字化社交媒体的快速发展和广泛普及，改变了人们获取时尚艺术信息和交流的方式。"2023时装艺术国际先锋展·中国濮院"将面向全球艺术家征集遴选的展览作品集结成册，以先锋艺术探索时尚文化与地域特色的交汇点，展现当代时装艺术与产业融合的无限可能。希望搭建起一个助力全球各地艺术家展示先锋时尚理念的互动平台，促进国际文化交流与合作，展现时装艺术多元性和包容性的态度。

图书在版编目（CIP）数据

介质律动：2023时装艺术国际先锋展·中国濮院/中国纺织工程学会，吕越主编.--北京：中国纺织出版社有限公司，2024.4

ISBN 978-7-5229-1517-3

Ⅰ.①介⋯ Ⅱ.①中⋯ ②吕⋯ Ⅲ.①服装设计—作品集—世界—现代 Ⅳ.①TS941.28

中国国家版本馆CIP数据核字（2024）第059205号

责任编辑：宗 静　　特约编辑：渠水清
责任校对：高 涵　　责任印制：王艳丽

中国纺织出版社有限公司出版发行
地址：北京市朝阳区百子湾东里A407号楼　邮政编码：100124
销售电话：010—67004422　传真：010—87155801
http://www.c-textilep.com
中国纺织出版社天猫旗舰店
官方微博http://weibo.com/2119887771
北京华联印刷有限公司印刷　各地新华书店经销
2024年4月第1版第1次印刷
开本：787×1092　1/16　印张：16.5
字数：300千字　定价：268.00元

凡购本书，如有缺页、倒页、脱页，由本社图书营销中心调换

编委会
Editorial Committee

主　编　吕越
Chief-editor: Lyu Yue（Aluna）

副 主 编　琴基淑（韩国）、李娟
Associate-editors: Key-Sook Geum (Korea), Li Juan

学术主持　许平
Academic advisor: Xu Ping

编 辑 组　罗杰、翟强、刘晓、赫然、张杨、杨颖、王欣源
Editing team: Luo Jie, Zhai Qiang, Liu Xiao, He Ran, Zhang Yang,
　　　　　　　Yang Ying, Wang Xinyuan

目 录

008 致辞与评论

008 时装艺术：对合作与团结的期待
琴基淑

012 立足于国际创新与时尚互动的时装艺术"介质律动"
——写在"2023 濮院时装艺术国际先锋展"开幕之际　许平

016 律动　吕越

021 时装艺术作品

022 安科·洛（美国/德国）《交织2》
024 鲍怿文《身体旅行》
026 贝纳兹·法拉希（美国）《虹彩》
028 毕然/姚智皓《记忆的形状》
030 蔡佳鹏《牵股》
032 陈艾《光影几何》
034 陈婵娟（美国）《点缀模块》
036 陈燕琳《布衣·流然》
038 成浩妍（韩国）《再现》
040 程澄《CHENGCHENG女孩的蝴蝶兰乌托邦》
042 邓鹤《同胚》
044 董磊《意识流——系列服装设计》
046 冯墨涵《制服》
048 赫然《记忆的形状》
050 洪伯明《劲象风华》
052 侯婷婷《浮沉》
054 胡虹慈/白紫千《Knitrusion》
056 胡霜叶《重重曲曲》
058 黄刚《文脉承续·中国染》
060 黄斯赞《水形颂》
062 晋长毅《错位》
064 凯瑟琳·冯·瑞星博（德国）《莨言》
066 袴着淳一（日本）《融合》
068 蓝星《未来图形主义》
070 劳达·拉瑞恩（美国）《纤维作画》
072 李莲姬（韩国）《寻找费尔岛》
074 李薇《空与影》
076 李迎军《毡衣无缝》
078 利昂·克雷赫廷（英国/俄罗斯）《谁掌握未来》
080 梁莉《方寸之间》
082 梁明玉《绝艳无色》
084 梁之茵《新衣》
086 刘辉《另一种生活》
088 刘静/姜绥祥（中国香港）《结1》
090 刘沁《虫洞》
092 刘薇《东临碣石　以观沧海》
094 刘晓天/杨达威《"身体"再定义：人机"共融"外骨骼设计》
096 刘欣珏（中国澳门）《祈愿娃娃们的疗愈艺术之旅》
098 刘鑫《反乌托邦》

100　刘寻《一根线＆青花》
102　刘伊童（澳大利亚）《脉冲》
104　刘伊童（澳大利亚）《澄澈流动》
106　卢禹君《线的流动与克制》
108　罗杰《少许春》
110　罗杰《新变量——麒麟》
112　罗莹《追忆》
114　吕越《百花深处》
116　梅丽莎·科尔曼/莱昂尼·斯梅尔特（荷兰）《震颤》
118　法瓦德·努里/苏维巴·法瓦德（巴基斯坦）《纪元天后：多元文化协同时尚律动》
120　穆芸《织衫织水》
122　潘瑶《山水空灵》
124　乔安娜·布拉特巴特（法国）《森林天使》
126　乔丹《斯威克时空基地》
128　秦耕《数字等身——虚拟服饰的现实映射》
130　琴基淑（韩国）《由点成线/由面到体》
132　邱辰《磁应》
134　莎拉·西维特（波兰/德国）《圆之貌》
136　莎拉·西维特（波兰/德国）《秋日旋律》
138　石历丽《花之可持续》
140　石梅《羽中曲》
142　苏杏《后人类主义启示录》
144　孙晓宇《无极》
146　唐芮《折叠城市》
148　涂淼淼《重塑》
150　王雷/李秋《取尘》
152　王文《心悦》
154　王欣源《公海里的阿芙罗狄忒》
156　王雪/苑国祥《流水廊桥》
158　王钰涛《和光同尘》
160　王悦《绩·续》
162　王志惠《繁花盛开》
164　吴帆/李频一《花器》
166　吴国禧（中国香港）/姜绶祥（中国香港）《虹霓经典》
168　吴晶《江南印象》
170　熊艺《2020年的天空》
172　徐秋宜（中国台湾）《奇幻奥德赛："浮生时尚"》
174　徐永鑫《似是故人来》
176　闫洪瑛《庆丰收》
178　严宜舒《掬月游》
180　杨敏《舞 舞 舞》
182　杨秋华《交·织》
184　叶锦添（中国香港）《手机》
186　余一萌《海骨》
188　余一萌/李舒祺/伍凝湘《平行意识》
190　庾晨溪《七日》
192　原一丹《昆虫幻想》
194　曾凤飞《际遇》
196　张刚《朦胧江南浓浓记》
198　张国云《琥珀蚕丝渡春秋》
200　张鹏《行走的中医》
202　张婷婷《青花记忆》
204　周朝晖《灵魂舞者》
206　周梦《盈盈一水间》
208　邹游《身体练习》
210　吕越/金小尧 桐乡市永欣服饰有限公司《日落时分》
212　赫然/浙江汇港时装有限公司《涌动》
214　鲍怿文/浙江浅秋针织服饰有限公司《衣作屏》

217　艺术家简历
253　艺术共创企业
260　组织机构

Contents

010 Addresses & Comments

010 Fashion Art: Anticipation of Collaboration and Solidarity   Key-Sook Geum

014 Fashion Art Based on International Innovation and Fashion Interaction, Rhythm of Media
—Written at the Opening of 2023 Puyuan International Fashion Art Invitational Exhibition   Xu Ping

018 The Rhythm   Lyu Yue(Aluna)

021 Fashion Art Works

022 Anke Loh(USA/Germany)   *INTERKNIT 2*
024 Bao Yiwen   *Let The Garment Carry Me*
026 Behnaz Farahi(USA)   *Iridescence*
028 Bi Ran/Yao Zhihao   *Shape of Memory*
030 Cai Jiapeng   *Holding Strands*
032 Chen Ai   *Geometry of light and shadow*
034 Chen Chanjuan(USA)   *Embellished Modularity*
036 Chen Yanlin   *Cloth·Flowing*
038 Chung Hoyeon(Korea)   *Repetition*
040 Cheng Cheng
CHENGCHENG Girl's Phalaenopsis Utopia
042 Deng He   *Homeomorphism*
044 Dong Lei
Stream of Consciousness–Series Clothing Design
046 Feng Mohan   *The Same Clothes*
048 He Ran   *Shape of Memory*
050 Hong Boming   *Elegance and Talent*
052 Hou Tingting   *Floating and Sinking*
054 Hu Hongci/Bai Ziqian   *Knitrusion*
056 Hu Shuangye   *Overlap*
058 Huang Gang   *Continuity of Context·Chinese Dye*
060 Huang Siyun
Kinetic Ode to the Underwater Wonderland
062 Jin Changyi   *Dislocation*
064 Kathrin von Rechenberg(Germany)   *Shuliang Talk*
066 Junichi Hakamaki(Japan)   *Fusion*
068 Lan Xing   *Future Graphicism*
070 Louda Larrain(USA)   *Painting with Fiber*
072 Lee Yeonhee(Korea)   *Finding Fair-Isle*
074 Li Wei   *Empty and Shadow*
076 Li Yingjun   *Seamless Felt Clothing*
078 Leonid Krykhtin(Britain/Russia)   *Who Hold the Future*
080 Liang Li   *Within a Square*
082 Liang Mingyu   *Colorless Stunning*
084 Liang Zhiyin   *New Dress*
086 Liu Hui   *Another Life*
088 Liu Jing/Kinor Jiang (Hongkong, China)   *Tied 1*
090 Liu Qin   *Wormhole*
092 Liu Wei
To View the Boundless Ocean from the Eastern Shore

094	Liu Xiaotian/Yang Dawei	156	Wang Xue/Yuan Guoxiang

094 Liu Xiaotian/Yang Dawei
Redefinition of "Body": Human-Machine "Co-Integration" Exoskeleton Design
096 Sanchia Lau(Macao, China)
Chiwawa: Country Road of Healing Art
098 Liu Xin   "N"topia
100 Liu Xun   A Thread & Blue and White
102 YT.LIU(Australia)   PULSE
104 YT.LIU(Australia)   Lucld Flux
106 Lu Yujun
The Flow and Restraint of the Line
108 Luo Jie   A Hint of Spring
110 Luo Jie   New Variable—Kylin
112 Luo Ying   Reminisce
114 Lyu Yue(Aluna)   Depth of Flowers
116 Melissa Coleman/Leonie Smelt(Netherlands)
Tremor
118 Fawad Noori/Suwaiba Fawad(Pakistan)
Epoch Diva: Rhythm of Mix culture through Fashion
120 Mu Yun
To weave with the legendary landscape
122 Pan Fan   Void Landscape
124 Johanna Braitbart(France)   Forest Angel
126 Qiao Dan   SIVICO Space base
128 Qin Geng
Digital Equivalents—a virtual mapping of the reality clothing
130 Key-Sook Geum(Korea)
Dots to Line/Shapes to Form
132 Qiu Chen   Magnetic Response
134 Sarah Siewert(Poland/Germany)
The Appearance of a Circle
136 Sarah Siewert(Poland/Germany)
Autumnal Melodies
138 Shi Lili   The Sustainability of Flowers
140 Shi Mei   A Song of the Feathers
142 Su Xing   Apocalypse of Post-humanism
144 Sun Xiaoyu   Wu Ji
146 Tang Rui   Folding Cities
148 Tu Miaomiao   Remolding
150 Wang Lei/Li Qiu   Collecting Dust
152 Wang Wen   Delight
154 Wang Xinyuan   Aphrodite of Open Sea

156 Wang Xue/Yuan Guoxiang
Harmony Bridge Over the Flowing Waters
158 Wang Yutao   Harmony of Light and Dust
160 Wang Yue   Ji·Continuation
162 Wang Zhihui   Flowers in Full Bloom
164 Wu Fan/Li Pinyi   The Vases
166 Haze Ng(HongKong, China)/
Kinor Jiang(HongKong, China)
Iridescent Classic
168 Wu Jing   Jiangnan Impression
170 Xiong Yi   the Sky in 2020
172 HSU CHIU I (Taiwan, China)
Fantasy Odyssey:"Fashion of the Floating World"
174 Xu Yongxin   As If the Old Friend Comes
176 Yan Hongying   Celebrate a Bumper Harvest
178 Yan Yishu   Water-Moon in Hands
180 Yang Min   Dance Dance Dance
182 Yang Qiuhua   Communication and Weaving
184 Tim Yip (HongKong, China)   Phone
186 Yu Yimeng   Sea Bones
188 Yu Yimeng/Li Shuqi/Wu Ningxiang
Parallel Consciousness
190 Yu Chenxi   7 Days
192 Yuan Yidan   Insects Fantasy
194 Zeng Fengfei   Encounter
196 Zhang Gang   Hazy Jiangnan thick memory
198 Zhang Guoyun   Amber Silk Spring and Autumn
200 Zhang Peng   Walking TCM
202 Zhang Tingting
The Memory of Blue and White Porcelain
204 Zhou Zhaohui   Soul Dancer
206 Zhou Meng   Between Flowing Waters
208 Zou You   Physical Exercises
210 Lyu Yue(Aluna)/Jin Xiaoyao/
Tongxiang Yongxin Clothing Co., Ltd.
Sunset Moment
212 He Ran/Zhejiang Huigang Fashion CO., Ltd.   Surge
214 Bao Yiwen/Zhejiang Qianqiu Knitwear Co., Ltd.
Clothes as Screen

217 Artists' Resumes

253 Co-create Enterprise

260 Organizations

致辞与评论

时装艺术：对合作与团结的期待
琴基淑

濮院，一座蕴含多元文化的古镇，针织行业的圣地，可以说是一个将科技、时尚、工业、艺术和谐相融的地方。通过在这座历史名镇举办本次时装艺术国际展，我希望能带来时尚与文化、艺术与科技、文化与纺织、时尚产业与国际性之间相互交流与合作的思考。

针织品最早出现在古埃及的文物中，而用于编织的针织机则是在16世纪被发明的。此后，针织衫主要作为工人的工作服或内衣。直到20世纪初，针织衬衫才被贵族作为高尔夫球衫穿着。针织品最早通过可可·香奈儿设计的平纹针织西装被引入时装设计中。香奈儿说："我借用平纹针织实现了舒适和自由，并创造了新的轮廓。"平纹针织西装已经成为创新的象征。香奈儿借用了既有的针织技术，展现了开创性的原创设计。

通过现代科学技术与时尚合作来完成的时装设计有着多种多样的案例。数字印刷的出现使表达各种色彩、织物纹理和图案成为可能，能更好地诠释文化和历史。与3D打印和激光切割等新兴现代技术的合作，也使原创时装设计成为可能。

艺术也是一个合作领域，它包含了人类在时尚中希望超越现实约束的愿望。时装艺术将艺术性最大化，允许人类的想象力自由表达其幻想。在这里，与现代科技的协作支撑了人类艺术想象力和幻想的实现。

特别是在20世纪，借助艺术与科学的积极协作，动态时尚和视频时尚等新时尚应运而生。移动或变形的时尚服装、发光的服装、变色的服装及采用可穿戴设备的服装，都是借助尖端技术的进步和协作来实现的。反射激光或全息图时装，以及机器人或人工智能的时装也是实现人类想象力的范例。

进入21世纪，现实与虚拟世界的界限已经消融，非现实的虚拟时装艺术也已崭露头角。元宇宙作为一种新的空间概念，正在逐渐成为真实空间的替代品。通过与时装艺术和时装设计领域的协作，快速变化的当代数字环境造就了新案例的诞生。通过这种方式，当代艺术与文化、技术和科学正在与时装艺术和时尚产业联手，相互交流影响，共同发展。这便是当代性或所谓的"时代精神"。

当代性在整个国际社会迅速传播，并通过交换新信息作为集体智慧变得无处不在。

世界的每个人都意识到国际合作和团结的价值。这是时装艺术需要国际同理心和协作的现实案例。

希望本次时装艺术展能进一步振兴濮院的针织技术和产业，也希望针织行业与时装艺术的交流与合作继续积极向前发展。

最后，我要由衷感谢濮院的众多组织人员和单位、机构以及 IFAN 组委会成员对本次时装艺术国际展的大力支持。我要向许多在幕后努力工作和帮助我的支持者表示衷心的感谢，同时也为国内外艺术家们提交艺术作品的激情热烈鼓掌。

亲爱的朋友们，让我们为 IFAN 喝彩！

琴基淑　教授

时装艺术国际同盟顾问
时装艺术国际同盟学术委员会主任
韩国弘益大学教授
韩国柳琴瓦当博物馆副主任
曾任韩国服装协会顾问、韩国时尚与文化协会顾问

Addresses&Comments

Fashion Art:
Anticipation of Collaboration and Solidarity
Key-Sook Geum

Puyuan, a city with diverse cultural heritage, is the Mecca of the knit industry and can be said to be a place where technology, fashion industry and art harmonize. By Opening this International Fashion Art Exhibition in historic city Puyuan, I hope to think about mutual exchange and cooperation between fashion and culture, art and technology, culture and textiles, and the fashion industry and internationality.

Knit were first discovered in ancient Egyptian artifacts, and the knitting machine for weaving knits was invented in the 16th century. After that, knits were mainly worn as work clothes or underwear for workers, and it was not until the early 20th century that knit shirts were worn as golf wear by aristocrats. Knit was accepted firstly into fashion design with Coco Chanel's jersey suit, and Chanel said, "I achieved comfort and freedom with knit jersey and created a new silhouette. " The jersey suit has become a symbol of innovation. Chanel presented innovative and original designs through collaboration with already existing knitting technology.

There are various examples of fashion design in which contemporary science and technology collaborated with fashion. The advent of digital printing has made it possible to express various colors, textures of fabrics, and patterns interpreting culture and history. Collaboration with emerging contemporary technologies such as 3D printing and laser cutting also made original fashion design possible.

Art is also a field of collaboration that embraces the human desire to transcend the limits of reality in fashion. Fashion Art, which maximizes artistry, allows human imagination to freely express fantasy. Here collaboration with contemporary science supports the realization of human artistic imagination and fantasy.

Especially in the 20th century, new fashions such as kinetic fashion and video fashion emerged due to active collaboration between art and science. Fashionable Clothing that moves or deforms, clothing that emits light, clothing that changes color, and clothing that incorporates wearable computers have all been made possible through the advancement of cutting-edge technology and collaboration. Fashion that reflects lasers or holograms, and fashion that utilizes robots or A are also examples of human imagination being realized.

In the 21st century, the boundary between reality and virtuality has collapsed, and unrealistic virtual Fashion Art has also appeared. Metaverse, which has emerged as a new concept of space, is also emerging as an alternative to real space. The rapidly changing contemporary digital environment is supporting the emergence of new cases through collaboration with the Fashion Art and fashion design fields. In this way, contemporary art and culture, technology and science are collaborating with Fashion Art and the fashion industry, exchanging influences and evolving each other. This is contemporaneity or the esprit of the times, "Zeitgeist".

Contemporaneity spreads rapidly throughout the international community and becomes omnipresent as collective intelligence through the exchange of new information. Everyone around the world realize the value of international cooperation and solidarity. This was presented as a realistic example of the international empathy and collaboration required in Fashion Art.

I hope that this Fashion Art Exhibition will further revitalize Puyuan's knitting technology and industry, and that exchanges and collaborations between knit industry and Fashion Art will continue in a positive way.

Lastly, I would like to express my sincere gratitude to the many Organizations and Companies in Puyuan, educational institutions, and organizing committee members of IFAN for supporting this fashion art exhibition possible. And I would like to express my sincere gratitude to the many supporters who worked hard and helped me behind the scenes. I also give a big round of applause to the passion of the artists who submitted their fashion art works at home and abroad.

Dear Friends, Bravo IFAN !

Key-Sook Geum, Ph.D.

Advisor, International Fashion Art Network
Director of the Academic Committee, International Fashion Art Network
Professor, Dept. of Textile Art Fashion Design, Hongik University
The Former Adviser, the Korea Society of Costume
The Former Adviser, the Korea Fashion&Culture Association

立足于国际创新与时尚互动的时装艺术"介质律动"
——写在"2023 濮院时装艺术国际先锋展"开幕之际

许平

2023 年的纺织时尚行业正在迎来新的发展阶段，作为全国时尚名城的浙江省桐乡市正积极抢占"未来产业"制高点。

为进一步加快濮院纺织服装产业向国际化、数字化转型升级，中国纺织工程学会与嘉兴市科学技术协会、桐乡市人民政府、濮院镇人民政府在浙江省嘉兴市濮院镇举办"介质律动——2023 濮院时装艺术国际先锋展"。本次展览希望由此搭建起一个助力全球各地艺术家展示先锋时尚理念的互动平台，立足于濮院悠久的文化历史和厚重的产业实力，实现时装艺术与跨媒体艺术、实验艺术、装置艺术、科技艺术、加密艺术等数字科技的创新联动，打造一场跨文化、跨媒介的时尚艺术大展，完成以"时装"为核心的时尚艺术综合表达，展现当代时装艺术与产业融合的无限可能。

本届展览主题"介质律动"，立足于开启国际创新与时尚互动的发展主题，力求运用全新的创作技巧，表达以时装为介质的协同艺术律动，折射出时装艺术家们对文明传承的反思以及对未来世界的探索。参展作品聚焦于科技赋能下的时装艺术，强调艺术创新对于设计创新的引领作用，紧密结合当地悠久多元的历史文化，从具体的作品创作元素和核心思想出发，按"产业互联""文化传承""时代征程""技术革新""数字联姻"五个方向，依托濮院这一方具有光荣革命传统的热土，挖掘红色资源、传承红色基因、弘扬红色经典，结合纺织时尚产业在当地融合发展的创新趋势，在新时代新经济新征程中讲好中国故事。

今天时尚与科技的关系正在进行着前所未有的转变，本次展览将力图在"十四五"新发展理念指引下，与国内外有志于此的艺术家一道，携手推进数字技术与经济、政治、文化、社会、生态文明建设"五位一体"的深度融合，以纺织时尚与艺术科技在新时代中的交叉赋能为总目标，探寻时装艺术之新、探究时尚生活之魅、探索纺织科技之未来。

许平　教授

国务院学位委员会第六届（艺术学）学科评议组委员、第七届（设计学）学科评议组召集人
北京设计学会会长
中央美术学院教授、博士生导师

Fashion Art Based on International Innovation and Fashion Interaction, Rhythm of Media
—Written at the Opening of 2023 Puyuan International Fashion Art Invitational Exhibition
Xu Ping

In 2023, the textile and fashion industry is ushering in a new stage of development, as the national fashion city, Tongxiang, Zhejiang Province, is actively seizing the heights of the "future industry".

In order to further accelerate the international and digital transformation of Puyuan's textile and garment industry, China Textile Engineering Society, Jiaxing Science and Technology Association, Tongxiang Municipal People's Government and Puyuan Town People's Government are scheduled to hold the 2023 Puyuan International Fashion Art Invitational Exhibition "Rhythm of Media" in Puyuan Town of Jiaxing City, Zhejiang Province. This exhibition hopes to build an interactive platform to help artists from all over the world to show their pioneering fashion concepts. Grounded in Puyuan's long cultural history and solid industrial strength, the exhibition seeks to achieve the innovative linkage between fashion art and cross-media art, experimental art, installation art, technology art, crypto art and other digital technologies, creating a cross-cultural and cross-media fashion art exhibition, completing the comprehensive expression of fashion art with "fashion" as its core, and demonstrating the infinite possibilities of the integration of contemporary fashion art and the industry.

The theme of this year's exhibition, "Rhythm of Media", is based on the development of international innovation and fashion interaction and seeks to apply new creative techniques to express the synergistic rhythms of art that uses fashion as a medium, reflecting fashion artists' rethink of cultural heritage and exploration of the future world. The exhibited works focus on fashion art under the empowerment of science and technology, emphasize the leading role of artistic innovation for design innovation, closely combine with the long and diversified history and culture of the location, and start from the specific creative elements and core ideas of the works; according to the five directions of "industrial interconnectivity" "cultural inheritance" "journey of the times" "technological innovation" and "digital connection".Relying on the Puyuan, a land with glorious revolutionary traditions, to excavate the red resources, they explore red genes, and promote the red classics. Combined with the innovation trend of the textile and fashion industry's integration and development at the local level, they tell a good story of China in the new journey of new economy in the new era.

The relationship between fashion and technology is undergoing unprecedented changes today. This exhibition aims to promote the deep integration of digital technology with economy, politics, culture, society and ecological civilization construction under the guidance of the new development concept of the "14th Five-Year Plan". Together with artists at home and abroad who are interested in this, the overall goal is to empower the intersection of textile fashion and art technology in the new era. We aim to explore the novelty of fashion art, investigate the charm of fashionable life, and explore the future of textile technology.

Xu Ping, Ph.D.

Member of the 6th municipal discipline appraisal(Arts)group and
Convenor of the 7th municipal discipline appraisal(Design)group,
Academic Degrees Committee of the State Council
Chairman, Beijing Design Society
Professor and doctoral supervisor, Central Academy of Fine Arts

今天，时尚与科技的关系正在进行着前所未有的转变，在世界数字技术的成果备受关注的时刻，我们策展的思路转向聚焦艺术与数字技术的结合。展览主题"介质律动"可以理解为：以时装为介质协同艺术律动，以科技为介质协同产业律动，以数字为介质协同视觉律动，以软材料为介质协同文化律动，以时尚为介质协同国际律动。这也是从 2007 年开始到 2023 年，时装艺术国际展已经走过的十七年间，我们一直在探索如何拓宽艺术边界、重新定义时装设计的一个延续动作。

本次展览立足于文化历史和产业支持，力求实现时装艺术与数字科技的创新联动。围绕历史文化精粹、邀约各国艺术家共同打造一场跨文化、跨媒介时尚艺术展览。以艺术与科技的视角对全新的创作材料和色彩、技巧和理念的表达，折射出时装艺术家们对文明传承的思索以及对未来世界前沿的探索。时装作为联结着人体和环境的纽带，承载的是人与世界交流的共情，随着全新的互动方式和生活体验不断开拓，现实和虚拟的局限得以打破，精神与物质世界的关系进一步融合，多元视野下的时装艺术让新时期的时尚身份得以重塑。

我们不断地拓宽边界，不断地尝试新技术、新材料的使用，增加新的观念。面对世界的变化，勇于表达感知感受，这也是大部分创作者从初期尝试到今天逐渐成熟作品的表现。

何为时装艺术？或许答案永远是动态的，因为我们展览中的作品一直在不断地刷新着大家的认知，迭代的、颠覆的、更新的。

吕越　教授

时装艺术国际同盟主席
时装艺术国际同盟学术委员会副主任
中国纺织工程学会时装艺术专业委员会主任
中央美术学院教授、博士生导师

The Rhythm
Lyu Yue (Aluna)

Today, the relationship between fashion and technology is undergoing an unprecedented transformation, and at a time when the fruits of the world's digital technologies are in the spotlight, our curatorial mindset is shifting to focus on the combination of art and digital technology. The theme of this exhibition "Rhythm of Media" can be understood as: Fashionable dress as a medium to synergize artistic rhythm, technology as a medium to synergize industrial rhythm, digital as a medium to synergize visual rhythm, soft materials as a medium to synergize cultural rhythm, fashion as a medium to synergize international rhythm. This is also a continuation of the action that we have been exploring how to broaden the boundaries of art and redefine fashion design in the 17 years that Fashion Art International Exhibition, which has been existing from 2007 to 2023. Based on cultural history and industrial support, this exhibition seeks to realize the innovative linkage between fashion art and digital technology.

Centered on the essence of history and culture, we invite artists from around the world to create a cross-cultural, cross-media fashion art exhibition. The expression of new creative materials and colors, techniques and concepts from the perspective of art and technology reflects the thoughts of fashion artists on the inheritance of civilization and their exploration of the future of the world's frontiers. Fashion, as a link between the human body and the environment, carries the empathetic communication between human beings and the world. With the continuous development of new ways of interaction and life experience, the limitations of reality and virtual reality can be broken, and the relationship between the spiritual and material worlds can be further fused, and fashion art under the diversified horizons allows the identity of fashion in the new era to be reshaped.

We are constantly expanding our boundaries, experimenting with new techniques and materials, and adding new concepts. Facing the changes in the world and expressing our feelings are also the manifestations of most of the creators' works that have matured from the initial attempts to today's mature works.

What is fashion art? Perhaps the answer is always changing, as the works in our exhibition are constantly refreshing, iterative, subversive, and renewing.

Lyu Yue (Aluna) , Ph.D.

Chairman, International Fashion Art Network
Deputy Director of the Academic Committee, International Fashion Art Network
Director, Fashion Art Professional Committee of China Textile Engineering Society
Professor and Doctoral Supervisor, Central Academy of Fine Arts

时装艺术作品
Fashion Art Works

Anke Loh (USA/Germany) 安科·洛（美国/德国）

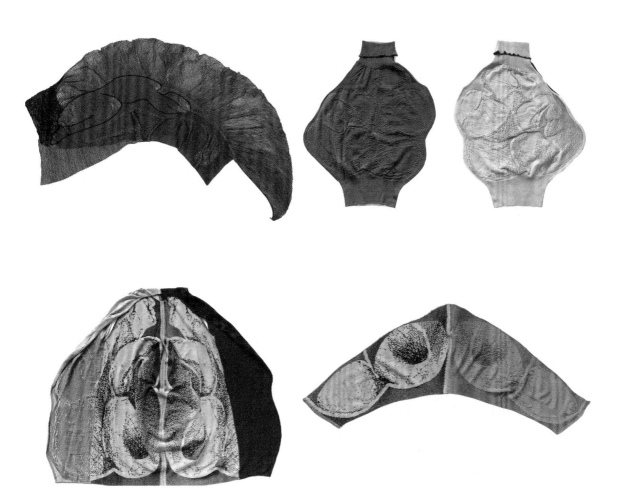

These abstract and unique pieces revolve around the female form. The collection of knits evokes women's stories—their opportunities, challenges, delights and anxieties. It's a knitting exploration that started with *Interknit* 1 during several residencies at the Textiellab in Tilburg, the Netherlands.

This concept collection emerged from my ongoing research on the female form and its significant place in culture, arts, and history. The collection includes sweaters, cardigans, scarves and sweater vests among other items.

本件作品由抽象而独特的针织片围绕着女性形体展开，这一系列的针织作品唤起了关于女性的故事——她们所面临的机遇、挑战、喜悦和焦虑。本系列的针织实验继续了创作者此前在荷兰蒂尔堡纺织实验室（Textiellab in Tilburg）对《交织1》系列作品的探索。

本系列的创作理念源于艺术家对女性形体及其在文化、艺术和历史中的重要地位的持续研究。该系列包括毛衣、开襟羊毛衫、围巾和毛背心等。

INTERKNIT 2
Merino wool

交织2
美利奴羊毛

Bao Yiwen 鲍怿文

Putting aside the traditional definitions of "body" and "clothing" to discuss their new definition, clothing can be a communication device between our body and the external space.
Using body elements to replace decorative elements in clothing, when there is no longer a clear division between the body and clothing, the concept of the body will be redefined, and the transformation from inside to outside is a "journey of the body".

抛开传统的"身体"与"服装"定义，讨论当下它们的全新定义，服装可以是我们身体与外界空间的沟通装置。利用身体元素置换服装中的装饰元素，当身体与服装不再有明确划分，身体的概念也将被重新定义，从内到外的转变是一场"身体的旅行"。

Let The Garment Carry Me
Mixed media

身体旅行
综合材料

Behnaz Farahi (USA) 贝纳兹·法拉希（美国）

What if our clothing could sense the movement and emotions of those around us? How might technology expand our sensory experience and influence our social interactions? And in what ways could our clothing become a form of non-verbal communication, expressed through changes in color and texture?

The hummingbird is a remarkable creature. The male Anna's hummingbird, for example, has feathers around his throat that appear at one moment completely green. With a twist of his head, however, he can turn them into an iridescent pink. He does this by exploiting the capacity of the microscopic structure of the feather to refract light like a prism, so that the feathers take on different shimmering hues, when viewed from different angles. This is how the Anna's hummingbird attracts mates during his spectacular displays of aerial courtship.

Iridescence is an interactive collar, inspired by the gorget of the Anna's hummingbird. It is equipped with a facial tracking camera and an array of 200 rotating quills. The custom-made quills flip their colors and start to make patterns, in response to the movement of onlookers and their facial expressions.

The goal is to explore how wearables can become not only a vehicle for self-expression, but also an extension of our sensory experience of the world. The advantage of using such an innovation can be to gather visual information such as people's facial expressions for those who have difficulties receiving or decoding this information such as those who suffers from visual impairment or autism. *Iridescence* can also express non-verbally and mimic those information, through its dynamic behavior. To make this work, we are borrowing from the latest advancement in AI facial expression tracking technology and embed it in bio-inspired material systems.Overall, this project is an attempt to explore the possibilities afforded by AI facial tracking technology and the dynamic behavior of a smart fashion item. The intention is to address psychosocial issues involving emotions and sensations, and to see how these technologies might inform social interaction.

如果我们的衣服能够感知周围人的动作和情绪会怎么样？技术如何扩展我们的感官体验并影响我们的社交互动？我们的服装如何成为一种非语言交流的形式，通过颜色和纹理的变化来表达？

蜂鸟是一种非凡的生物。例如，雄性安娜蜂鸟喉咙周围的羽毛一度完全呈现绿色。然而，只要转动它的头，就能将羽毛变成虹彩般的粉红色。利用羽毛微观结构的能力，像棱镜一样折射光线，使羽毛在不同角度观察时呈现出不同的微光。这就是安娜蜂鸟在空中求爱表演时吸引配偶的方式。《虹彩》是一款交互项圈，其灵感源自安娜蜂鸟的颈围。它配备了一个面部跟踪摄像头和一个由200个旋转套筒组成的阵列。定制的羽毛片根据围观者的动作和面部表情，翻转颜色并开始制作图案。

作品的目标是探索可穿戴设备如何不但成为自我表达的工具，而且成为我们对世界感官体验的延伸。使用这种创新的优点是可以为那些难以接收或解码这些信息的人（如患有视力障碍或孤独症的人）以收集视觉信息，如人们的面部表情。《虹彩》还可以通过其动态行为，以非语言方式表达并模仿这些信息。为了实现这一目标，我们借鉴了人工智能面部表情跟踪技术的最新进展，并将其嵌入到仿生材料系统中。总的来说，该项目旨在探索人工智能面部跟踪技术，和为智能时尚产品的动态行为所提供的可能性。其目的是解决涉及情绪和感觉的社会心理问题，并了解这些技术如何影响社交互动。

Iridescence
3D printed, electromagnetic actuators, electronics

虹彩
3D 打印、电磁制动器、电子电路

Bi Ran/Yao Zhihao 毕然/姚智皓

The work is inspired by the contemporary interpretation of traditional Chinese culture, depicting oriental cultural forms in memory through new technological means, and tapping into and inheriting the charm of traditional culture.
The work adopts a future Chinese style, using a multidisciplinary cross-border fusion of design vocabulary to present the Chinese flavour, integrating traditional Chinese culture with international aesthetics, so as to form a contemporary and forward-looking design expression with Chinese cultural heritage.
The costume works are completed by 3D printing and laser cutting technology, and the video works are completed by motion capture, 3D scanning technology and Blender modelling. The visual aesthetics of the fusion of technology and culture is presented in contemporary terms, linking the relationship between history, technology and the future.

作品灵感源于对于中国传统文化的当代解读，通过新技术手段描绘记忆中的东方文化形态，挖掘、继承传统文化的神韵。
作品采用未来中式风格，运用多学科跨界融合的设计语汇呈现中国韵味，将中国传统文化与国际化审美相融合，以此形成具有中国文化底蕴的当代前瞻性设计表达。
服装作品整体由 3D 打印、激光切割技术完成，视频作品使用动作捕捉、3D 扫描等技术、Blender 建模制作完成。用当代语汇呈现科技与文化相融合的视觉美感，串联出历史、科技和未来的关系。

Shape of Memory
Plastic laser cutting 3D printing

记忆的形状
塑料激光切割 3D 打印

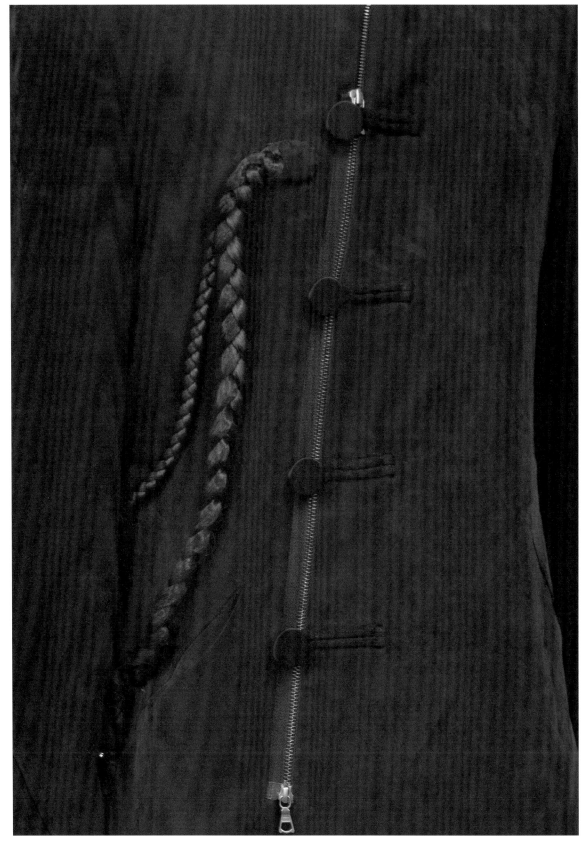

The creative motivation stems from the strong sense of inheritance and integration that the development of Chinese and Western cultures has brought to the artist to this day. On the basis of traditional Chinese clothing, a design approach with contemporary Western avant-garde cutting characteristics has been integrated.

Based on the physical characteristics of Chinese people, improvements and optimizations have been made in the silhouette and the relationship between the sleeve and the body; At the same time, it creates a contrast between ancient and modern, Chinese and foreign cultures in multiple dimensions of space, reflecting the mutual influence and integration of Chinese and Western cultures.

The color was chosen from the traditional blue color of ancient China; The right front panel is connected with a high-temperature silk woven oxtail braid, extending to the side of the body, aiming to emphasize the inheritance and continuity of cultural heritage based on the current multicultural environment.

The artist, using fashion as a medium, attempts to combine traditional culture with innovation and expansion, which is a pioneering attempt at culture and attitude, technology and art.

创作动机来源于中西文化发展至今给作者带来的强烈传承感与交融感。该作品在传统中华服饰的基础上融合了具有当代西方先锋剪裁特征的设计方式。

该作品依据中国人的身体特征在廓型与袖身关系上做出了改良与优化；同时在空间的多个维度上营造了古今中外等层面的对照，体现出中西文化的相互影响和融合。颜色选用了中国古代传统的青色；右前片接有高温丝编织而成的牛尾辫，延伸至身体侧面，意在强调立足于当前多元文化组成的大环境下，文脉的传承与绵延不绝。

作者以时装为媒介，尝试将传统文化与创新拓展相结合，是一次对于文化与态度、技术与艺术的先锋尝试。

Holding Strands
Cupro, high temperature fiber (digital printing)

牵股
铜氨丝、高温丝（数码印）

Digital Virtual Fashion. Light and Shadow, Linear surface structure, repetition, transparency, lamination, light effect.

数字虚拟时尚，光影，几何线面结构，重复，透明，层叠，光效。

Geometry of light and shadow
Virtual works/Video format

光影几何
虚拟作品 / 视频形式

Chen Chanjuan (USA) 陈婵娟（美国）

The inspiration for this design originated from traditional knitting and embroidery techniques. By seamlessly merging a 3D printer and a 3D pen into one design using the thermoplastic polyurethane (TPU) filament, the artist combined the precision of computing and digital fabrication offered by 3D printers, with the tactile artistry of a craftsman wielding a 3D pen technique, to create unexpected fabric-like textiles. The motifs used in this design is a floral and botanical theme knitted together, representing the beauty of the traditional textile making process. The combination of 3D printing and 3D pen represents creative intelligence and skill wedded together. In addition, the modular concept eliminated the use of sewing and allowed the design to be customizable by altering the placement of the modular hoops.

此设计的灵感源自传统的针织和刺绣技术。艺术家利用热塑性聚氨酯（TPU）长丝将 3D 打印机和 3D 笔无缝地融合到一个设计中，将 3D 打印机提供的计算和数字制造的精确性与手握 3D 笔技术的工匠的触觉艺术性结合起来，创造出出乎意料的类似织物的纺织品。此设计采用的图案是由花卉和植物主题编织在一起的，这代表了传统纺织制作过程的美感。3D 打印和 3D 笔的结合代表了创新智能和技能的结合。此外，模块化的概念消除了缝纫的使用，并允许通过改变模块环的位置来定制设计。

Embellished Modularity
3D printed thermoplastic polyurethane (TPU) filament

点缀模块
3D 打印的热塑性聚氨酯（TPU）长丝

Chen Yanlin 陈燕琳

The design inspiration comes from parametric architectural modeling, and transforms its nonlinear form into fashion art creation. The most original cotton grey cloth and the oldest pleating technology are used to create a free-growing flowing fashion art effect, and explore the new value of dynamic fashion aesthetics that moves between static and dynamic. At the same time, the rustic materials and the curved space of organic form are interwoven to evoke people's desire for closeness to nature, showing the affinity between man and nature, while extending a kind of endless visual tension, full of vitality, returning to the original, containing a sense of life, and reinterpreting the contemporary significance of fashion.

设计灵感源于参数化建筑造型，将其非线性形态转化到时装艺术创作中，用最原始的棉坯布、最为古老的百褶工艺，塑造出自由生长的流动感时装艺术效果，发掘游走在动静之间的时装美学新价值。同时，将质朴材料与有机形态的曲线空间交织一体，唤起人们对自然亲近的渴望，表现出人与自然的亲和美感，同时延伸出一种生生不息的视觉张力，充满生机活力，回归原始，蕴含着生命感，重新诠释时尚的当代意义。

Cloth · Flowing
Cotton

布衣 · 流然
棉布

Chung Hoyeon (Korea) 成浩妍（韩国）

Title *Repetion* it's body of work highlights permeable material, constructing organic shapes with polyester mesh. Each work contains a record of the artist's choices in its construction, including brush strokes of color and serged edges. Both resilient and delicate, voluminous but airy, the material allows Chung to allude to thoughts, memory, and inspirations beyond the tangible.

题目为《再现》(*Repitation*)的作品集强调了可渗透的材料，用聚酯网状材料构造出有机形状。每个作品都包含了艺术家在其构造中的选择记录，包括色彩的笔触和包边的边缘。这种材料既坚韧又精致，既丰满又轻盈，使作者能够引申出超越实物的思考、记忆和灵感。

Repetition
polyester mesh

再现
聚酯纤维网格材料

The video tells the story of CHENGCHENG girl's Phalaenopsis utopia.
Butterfly orchids fly into CHENG CHENG's Phalaenopsis girl Utopia, bringing good luck to the world. The Phalaenopsis girls here live in this wonderful world wearing dresses made of flowers and water. They believe that flowers bring good luck and love flowers and nature.
Phalaenopsis flowers bring you into this beautiful and romantic world of flowers.
Creative Team:
The Stylist: ChengCheng
Creative design: Cheng Cheng, Yin Hongchen
Project Coordinator: Liu Yanjia, Li Chengyu
Space creativity: Zhang Shupeng, Ding Ding, Wei Luning, Zhou Jie, Zhou Yanshan
Design Coordinator: Chen Ziqian, Wei Luning
3D vision: Zhang Yu, Liu Jianxiang and Shen Shuting
Topology map: Wang Hao, Zhang Fu Wei, Huang Lei, Ming Xie Jin
Dynamic effect design: Fu Haochen, Ji Wenjun, Liu Jianxiang, Chen Jiawei, Luan Zhe, Zhu Lingyun, He Yanglu, Wu Yanran
UE5 Engine: Chen Xu, Yang Qihui, Liu Wang, Zhang Yu, Liu Jianxiang, Shen Shuting
Video clip: Wei Luning, Ji Wenjun, Dang Junhao
SKU Design: Qi Haoxiang, Liu Shenghao, Pan Junjie, Sun Xiaoshuai

视频里讲述了 CHENGCHENG 女孩的蝴蝶兰乌托邦。

蝴蝶兰花化身精灵飞入 CHENG CHENG 的蝴蝶兰女孩乌托邦，将好运带给了这个世界。在这里的蝴蝶兰女孩们穿着由花与水构成的裙子，生活在这个奇妙的世界。她们相信花能给人带来好运，并热爱着花朵与大自然。

蝴蝶兰花带你走进美丽浪漫的花朵世界。

创作团队：
设计师：CHENGCHENG
创意设计：程澄　尹宏晨
项目统筹：刘艳佳　李程昱
空间创意：张树鹏　丁丁　魏鲁宁　周洁　周琰姗
设计统筹：陈子谦　魏鲁宁
3D 视觉：张宇　刘建祥　沈舒婷
拓扑贴图：王昊　张付伟　黄磊　代明　谢金
动效设计：付浩辰　吉文军　刘建祥　陈嘉伟　栾哲　朱凌云　贺洋璐　吴延然
UE5 引擎：陈旭　杨淇惠　刘旺　张宇　刘建祥　沈舒婷
视频剪辑：魏鲁宁　吉文军　党俊豪
SKU 设计：齐浩祥　刘圣浩　潘俊杰　孙晓帅

CHENGCHENG Girl's Phalaenopsis Utopia
Virtual works/Video format

CHENGCHENG 女孩的蝴蝶兰乌托邦
虚拟作品 / 视频形式

Deng He 邓鹤

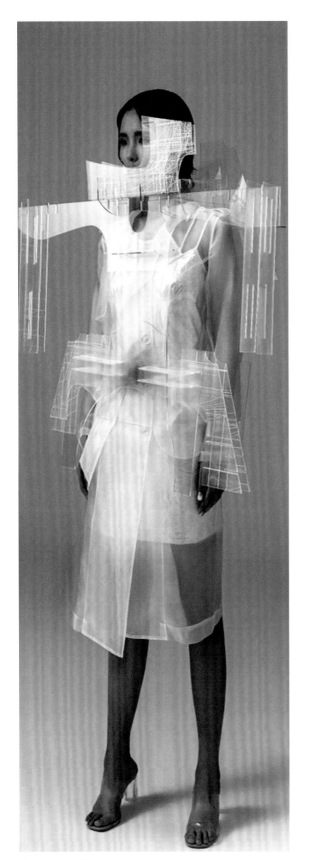

The tolerance and modesty of traditional Chinese culture is reflected in the subtle changes of the lines in traditional clothing cut-outs. The work juxtaposes the two-dimensional cut-outs of traditional clothing from different periods with modern clothing, reconstructing a brand-new image to show a new perspective on the inheritance and innovation of traditional culture.

中国传统文化的包容并蓄与谦逊体现在传统服装裁片线条微妙的变化之中，将不同时期传统服装二维裁片与现代服装并置，重构全新的形象，展现当代传统文化传承创新的新角度。

Homeomorphism
Mixed media

同胚
综合材料

Dong Lei 董磊

The *Stream of consciousness* series is designed to establish a stream of consciousness between the objective real world and the subjective consciousness world, which is an interpretation of human consciousness.Through rational consciousness and irrational and illogical subconscious, objective and subjective feelings are intuitively explored, swinging between cognition and unknown.

The process and observation of using a series of dreamy colors and structures to form a compound sense and a sense of life are similar to a continuous flow of water, and the image of clothing is the gradually rising waves of this flow, and the irrational intuition and dream nerve collide with the rational consciousness field.

The possibility of experimenting with non-textile materials in the selection of materials highlights the magic and agility of the future. The material selection is bold enough to break through the plain situation of ordinary fabrics and give people a refreshing feeling in clothing. Breaking the inherent definition of clothing design and the dividing line of different material properties brings a new possibility to clothing design.

《意识流》系列设计是在客观现实世界与主观意识世界之间建立的意识流，是关于人类意识的阐释，通过理性意识和非理性无逻辑的潜意识对客观和主观感觉进行直观探索，在认知和未知之间摆动。

该设计用一系列梦幻般的色彩、结构形成一种复合感、生活感的过程和观察，类似于一条切不断的流水，而服装的意象是这股流水溅起的浪花，用非理性的直觉、梦幻神经碰撞理性的意识领域。

该设计在选料上多多尝试非纺织材料的可能性突出未来的魔幻、灵动。选料足够大胆，突破普通面料平淡无华的处境，在服饰上给人耳目一新的感觉。打破了服装设计固有的定义和不同材料使用性质的分界线，为服装设计带来一种新的可能性。

Stream of Consciousness–Series Clothing Design
PP (Polypropylene) nylon fiber elastic, wool spinning

意识流——系列服装设计
PP（聚丙烯）尼龙纤维弹丝、羊毛纺纱

Feng Mohan 冯墨涵

Evolution is a change without direction, it can be an evolution from simple to complex or a degradation from complex to simple. The work starts from the evolution of plants and tries to analyze the mutual transformation of complexity and simplicity in the environmental change and get regularity revelation from it. In the long river of history, clothing, as a body wear, has also experienced a long evolution, *The Same Clothes* is based on the law of evolution, focusing on the predictable future under the development of digital technology, to explore the future evolution of clothing. With the continuous development and maturity of technologies such as glass-free 3D, the boundary between the virtual and the real has been blurred. The visual and aesthetic needs of "complexity" in clothing will be replaced by virtual clothing, while the physiological and functional needs of "simplicity" in clothing will continue to be undertaken by physical clothing. When the evolution of "complexity" and "simplicity" is complete, at some point in the future, we may all be wearing the same "simple" clothes in real life. The artist explores the possible evolution of clothing under the intervention of digital technology with speculative thinking and finds a similar relationship between the symbolic role of clothing and the symbolic metaphor of plants. Under the theme of cultural identity, the study of plant forms and specific patterns present an exploration of "complexity".

演化是没有方向的变化，可以是由简单到复杂的进化，也可以是由复杂到简单的退化。作品从植物的演化展开研究，试图分析环境变化中繁与简的相互转化，从中得到规律性的启示。在历史的长河中，服装作为身体穿戴物，也经历了漫长的演变，*The Same Clothes* 基于演化的规律，将视角聚焦于数字技术发展之下的可预测未来，来探讨服装的未来演化。随着 3D 裸眼等技术的不断发展与成熟，虚拟与现实的边界变得模糊。服装中"繁"的视觉与审美需求将由虚拟服装替代执行，而服装中"简"的生理与功能需求将由实体服装继续承担。当"繁"与"简"完成进化，在未来的某个情景下，现实中的我们可能都穿着同样至"简"的服装。作者以思辨的方式探索了数字技术介入下服装脉络演化的可能，并找到服装作为象征作用与植物象征隐喻的类似关系，在文化身份的主题下，对植物形态与具体版型的研究呈现对于"繁"的探索。

The Same Clothes
Virtual works/Video format

制服
虚拟作品 / 视频形式

He Ran 赫然

The *Shape of Memory* starts from the perspective of the fashion show and initially focuses on the Unwearable Art in fashion art, concentrating on exploring the possibility of clothing as an art medium.

Inspired by my concerns and thoughts about the flux of my own growing path, personal and collective identity, memory and shape are opposites, the contradiction between the abstract and the figurative, experiences and emotions memory carries, have long been condensed into history and culture, and influence the development of the individual. I place the perspective of the city where I grew up in my childhood and magnify the unique ice crystal elements of the northern city, the unique natural climate and environment, to give it a thousand fissures. The unique natural climate and environment of the city gives it a variety of fissured shapes and uncertainties and maps the diversity of individual and group growth and development. Depicting the natural environment from a microscopic point of view, the conceptual elements of ice crystals, glaciers, ice sculptures, plants, light and shadow, etc. are used to portray abstract memories into concrete shapes with the use of the body.

The carrier of "unwearable clothing" in the show becomes a comprehensive art form that combines artistic concepts, experimental materials, and non-functional needs.

Shape of Memory
Mixed media

作品《记忆的形状》从秀场视角入手，在研究中首先聚焦于时装艺术中的不可穿戴的艺术，专注于探索服装作为一种艺术媒介的可能性。

灵感源于我对自身成长路径流变、个人与集体的身份认同等问题的关注与思考，记忆和形状本就是对立的，抽象与具象的矛盾，记忆承载着经历、情感，长久以来凝结成历史与文化，并影响着个人的发展。我将视角置于童年成长的城市之中，将北方城市独有的冰晶元素放大，特有的自然气候与环境给予其千姿百态的裂变形状和不确定性，也映射了个人与群体成长与发展的多样性。描摹微观视角下的自然环境，用冰晶、冰河、冰雕、植物、光影等为概念元素，借用身体将抽象的记忆描摹出具体的形状。

在秀场中所承担"不可穿服装"的载体就变成了集艺术性概念、实验性材料、非功能性需求等多要素于一体的综合艺术形式。

记忆的形状
综合材料

Hong Boming 洪伯明

Integrating from traditional Chinese utensil culture and spirit of ingenuity, taking the spirit of Chinese jackets which rooted daily-wear and intangible cultural heritage. Combining the inspiration from utensil culture and the new ways of ancient arts (shadow puppets, digital printing). Shadow puppetry is a traditional art that combines exquisite carving technology and melodious stories, same as K-BOXING(cultural heritage, carefully selected materials, and craftsmanship), investing the brand with humanistic and artistic expression of heritage and ingenuity, interpreting the aesthetic philosophy. The shape of dragon as convey meaning, demonstrating the mellow accumulation and international vision of the "new high-end domestic brand".

该设计融器物之道，从醇厚的中国传统文化与匠心精神基底出发，取中国茄克根植于日用之道与非物质文化遗产的薪火相承气魄，合器物共生、古艺新用（皮影、科技数码印花）之灵感。皮影是通过精湛的人物雕刻剪影配以曲调故事表演的民间传统艺术，取皮影与劲霸的共性（文化传承、臻选材质、匠心匠艺），赋予劲霸茄克传承匠心的人文艺术表达，演绎劲霸独特的东方器物美学哲思。

同时，该设计以龙化形，以形传意，彰显"高端新国货"品牌的醇厚积淀与国际视野。

Elegance and Talent
Mixed media

劲象风华
综合材料

The inspiration initially came from the cellular structure of plant tissue observed under a microscope, and adhered to the original intention of multi-dimensional inheritance of intangible cultural heritage technology.
The design adopts asymmetric deconstruction design, relying on the patterned texture of bamboo weaving. The top fabric is mainly hand-knitted with jet yarn, wool, cotton yarn, and fish thread, while the lower skirt is made of vertical double crepe silver chiffon yarn and tencel cotton linen fabric. Traditional grass and wood dyeing and spray dyeing processes are used, and the design uses manual knitting to reproduce the texture of Huazhou intangible cultural heritage bamboo weaving. While activating the new vitality of traditional handicrafts, Also expanding and creating materials for knitted fabric design. The work explores the essential connection between humans and natural organisms, with the aim of returning to the essence of nature.

该设计灵感最初来源于在显微镜下观察到的植物组织细胞结构，以及秉承着多维度传承非遗工艺的初心。
设计廓型选用不对称解构设计，以竹编的花型肌理为依托，上衣面料用喷毛纱、羊毛、棉毛线和鱼线手工针织为主，下裙为竖纹双绉银丝雪纺纱和天丝棉麻面料，采用传统草木染与喷染工艺。设计运用手工针织重现华州非遗竹编编织肌理，在激活传统手工艺新生命力的同时，也为针织面料设计拓展创作素材。
作品探究人与自然生物间的本质联系，意在回归自然本体。

Floating and Sinking
Mixed media

浮沉
综合材料

Does technology impose limitations upon humanity?
Does it bring boundless inspiration within confined spaces?
Is it breaking through dimensions or endlessly extending within the same dimension?
Based on contemplation of these questions, Hongci has created a series of textile sculptures rooted in the art of knitting, exploring the possibilities that textile technology bestows upon humanity. The piece *Knitrusion* takes its inspiration from the realm of modelling, where "Extrusion" signifies the technique of transforming surfaces into three-dimensional forms, and "Knit" refers to the media utilized: knitting.
The artwork takes the form of a knitted bodysuit, elongating of the human form through extending the shoulders to an extraordinary height. Within a space of nearly four meters, individuals are granted the freedom to engage in limitless actions and movements. The elongated garment and the diminutive figure form together serve as a metaphor, perhaps interpreting the possibilities that technology presents. Is it boundless? Or boundless confined within limits? Are we the figure within the sculpture narrative?
"Knitrusion" is supported by Huading Sports Equipment Co., Ltd., providing technical expertise.

科技是否给人类带来限制？
它在有限的空间内带来无限的灵感？
它在突破维度吗？还是在同一维度下的无尽延伸？
基于对这些问题的思考，作者创作了一系列以针织为基底的纺织雕塑作品，探索纺织科技给人类带来的可能性。
作品 *Knitrusion* 是以建模的设计方式为灵感进行构思的，"Extrusion" 指建模中 "将面转化为体" 的技术，"Knit" 指作品采用的媒介——针织。
作品以针织连体紧身衣为雏形，通过拉长肩部将人体延伸至一个非正常高度。在近 4 米的空间里，人在既定的范围内可以做出无限的动作和行为。拉长的服装和渺小的人体，这个行为艺术似乎在解释科技带来的可能性——它是无限制的？还是有限的无限制？我们是否就是雕塑中的那个人？
本作品由桦鼎运动用品有限公司提供技术支持。

Knitrusion
Knitted brocade, spandex fabric, chloroprene rubber, nylon

Knitrusion
针织锦、氨纶面料、氯丁橡胶、尼龙

What will clothing look like in the metaverse? Should it be like clothing in reality, with many inspirations coming from nature and ecology? Does it still retain the essence of clothing being worn? Does it still maintain the essence of clothing design that continues to explore in terms of silhouette, color, and materials? Yes, it seems not.

Digital works starting from environmental simulation, inspired by the reconstructed nature, attempt to use real-life design logic for virtual clothing design. Using digital technology to simulate the combination of fabrics and hard materials that cannot be achieved in the real world, to explore design concepts between virtual and real. The words' heavy and picturesque, and the melody like a screen 'complement each other in terms of reality and emptiness.

Overlap
Virtual works/Video format

在元宇宙中的服装会是什么样子？它应该是和现实中的服装一样，很多灵感来源于自然与生态吗？它依然保有服装被穿戴的本质吗？它依然保有服装设计在廓型、色彩、材料上继续探索的本质吗？是，又好像不是。
从环境模拟着手的数字作品，以重新构建的自然为灵感来源，尝试使用现实中的设计逻辑来进行虚拟服装的设计。运用数字技术模拟不能在现实世界组合的面料与硬质材料，探索虚实之间的设计理念。"重重如画，曲曲如屏"，虚实相生。

重重曲曲
虚拟作品 / 视频形式

The designer of this series combines traditional Chinese dyeing techniques such as tie-dyeing and hanging dyeing with modern digital printing and 3D printing knitting technology. Bio-based fibers and GRS certified recycled and environmentally friendly fibers are used in fabrics, and out-of-season stock yarns redesigned by designers are used in local decoration of garment design to express the design concept of integrating tradition and modernity and sustainable development.

本系列作品，设计师将扎染、吊染等中国传统染色工艺与现代数码印花、3D 打印针织技术相结合。在面料上多选用生物基纤维、GRS 认证再生环保纤维，经设计师重新设计的过季库存纱线用于成衣设计的局部装饰，以表达传统与现代相融合及可持续发展的设计理念。

Continuity of Context · Chinese Dye
Hemp, bio-based fiber, GRS certified regenerative fiber

文脉承续 · 中国染
汉麻、生物基纤维、GRS 认证再生环保纤维

Kinetic Ode to the Underwater Wonderland is a collection of kinetic & interactive fashion art creation. "Perceptible, interactive and morphable", by using shape-changing elements to convey the fluidity and beauty of the underwater world. It combines the knowledge and methods of fashion art, speculative design, human-computer interaction and bio-inspired kinetic structure, etc. Multiple sensors, dynamic materials and devices are embedded and can react to the changing environment.

The material adopts transparent acrylic with special coating, and gradient blue mesh, which fluctuates in light and shadow, movement and space, like breathing, and growth, it tend to describe the aesthetic concept of artificial order and natural spirituality. The series is presented in a water-light space that simulates sea level, creating an immersive multi-sensory perception and experience, and composing the sense of responsibility and sustainability implied in the relationship between human and ecology.

《水形颂》是一件动态时装艺术作品。"可感知的、可响应式的和可变形的"通过使用形态变化的元素来展现海底世界的流动性和生命力。作品集合时装艺术、思辨设计、人机交互、仿生动力学结构等跨领域思维与工具，在时装中嵌入传感器与动态材料、装置，使其对不断变化的外界环境做出反应。

材料采用附着特殊涂层的透明亚克力、轻柔的渐变网纱，在光影、运动和空间中起伏波动，似呼吸、似生长，叙述人工秩序与自然灵性的美学理念。该系列模拟海平面下的水光空间，营造沉浸式的多感官感知与体验，谱写人与生态关系间隐含的责任感与可持续性。

Kinetic Ode to the Underwater Wonderland
Organza, acrylic, comprehensive material

水形颂
欧根纱、亚克力、综合材料

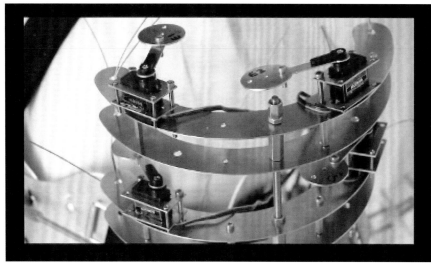

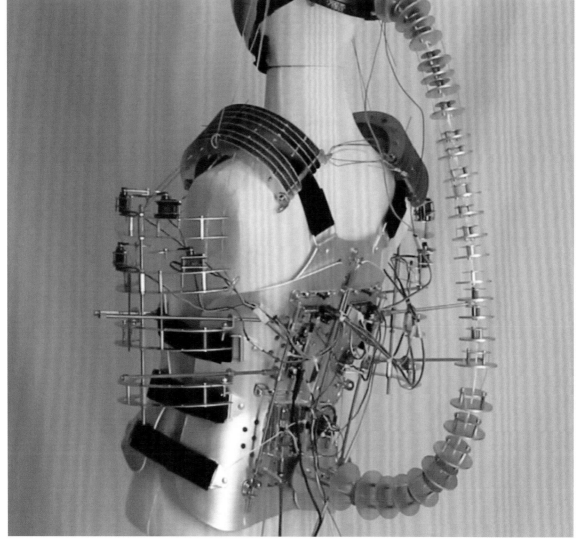

Jin Changyi 晋长毅

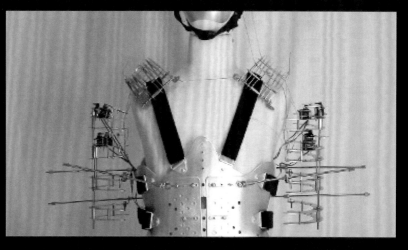

Behind the perceptual experience lies the hidden emotions of the human heart, among which the sense of touch is the emotional memory with strong universality. The dislocation of touch is also the dislocation of subjectivity.

The work aims to create the experience of tactile dislocation between two people. The work consists of two parts: two people in different spaces, and the tactile sensing devices worn by each of them. They are in different spaces, but their smart wearable devices interact remotely with each other based on changes in body data, allowing them to feel each other's tactile experience without touching, and to use their own bodies to perceive the experience of others. The armor in *Dislocation* loses its original defensive function and turns into an invasive torture device. The reproduction, experience and understanding of bodily pain enables individuals in different spaces to realize emotional connection, and the wearer can be more clearly aware of his own existence while sensing the tactile experience of others.

知觉经验的背后蕴含着人们内心隐秘的情感，其中，触觉是具备较强普适性的情感记忆。触觉的错位，也是主体性的错位。

作品旨在营造两个人触觉错位的体验，包含两部分：身处不同空间的两人，各自穿戴的触觉感应设备。二人身处不同的空间，但他们的智能可穿戴设备会根据身体数据变化而产生远程互动，令他们无须触摸，就能够感受到彼此的触觉体验，借助自己身体去感知他人经历。《错位》中的盔甲失去原有的防卫功能，转为入侵式的刑具，对身体痛感的再现、体会、理解，使身处不同空间的个体实现情感连接。穿戴者在感知他人触知体验的同时，可更为清晰地意识到自身生命的存在。

Dislocation
Electronic components, metal, acrylic, screen

错位
电子元件、金属、亚克力、屏幕

Kathrin von Rechenberg (Germany) 凯瑟琳·冯·瑞星博（德国）

Its unpredictable colors, the energy of the sun, and the flowing form of the fabric are in rhythm with nature. It is born in nature and returns to nature. The art work *Shuliang Talk* composed of three parts: Plain white silk shibori technique degummed dress, soft sculpture made of silk with Shuliang nature dyed and Shuliang hand dyed evening dress. Plain white silk is what it looked like before it became the Xiangyusnha. China is the root country of silk, and its natural and pure beauty always makes Kathrin's heart flutter. Inspired by the traditional way of Xiangyunsha, Kathrin repeatedly brushes the surface of naturally dyed silk with Shuliang juice and leaves them under the sun to give the fabrics stiffness and adds "confidence" to them.

《莨言》是用真丝与薯莨染色制作而成。它变幻莫测的颜色、吸收阳光的能量、面料自由的形态，是自然的律动。其本生于自然，终归于自然。作品由素白真丝脱胶雕塑裙、薯莨染色真丝软雕塑、薯莨染色裙三部分组成。素白真丝是成为香云纱前"素颜"的样子。中国是丝绸的发源地，它天然纯净的美，总能让凯瑟琳为之心动。受香云纱传统染色方式的启发，凯瑟琳将天然染色的真丝面料表面反复刷薯莨汁，日晒赋予面料硬度，为面料增加"底气"。

Shuliang Talk
Silk

莨言
真丝

Junichi Hakamaki (Japan)　　　　　　　　　袴着淳一（日本）

The design inspiration of this work comes from the fusion of Eastern and Western cultures, combining Western style suits with the waistband of Japanese kimonos in style expression. And in some parts of the suit, such as the wide sleeve shape, kimono elements were also implanted, but the classic cut of the Western style suit was chosen, which represents the exquisite and elegant fusion of Eastern and Western cultures. The waistband element of Japanese kimono has been added to the waist, which is not only a practical clothing accessory, but also strengthens the way women are shaped in Western clothing and pays tribute to traditional Japanese clothing. The material and color of the belt also fully reflect the delicacy and delicacy of Japanese culture. In terms of fabric selection, traditional Japanese fan shaped patterns and floral patterns were selected, combining embroidery and printing, reflecting both the texture of Western fashion and showcasing the magnificence of Eastern fabrics. I believe that design should represent a cultural exchange and integration, presenting a wonderful intersection of Eastern and Western cultures. This work is not only fashionable, but also a novel expression of culture, showcasing the diversity and beauty of today's fashion, art, and other cultures.

Fusion
Textile fabrics

该作品的设计灵感来源于东西方文化的融合，在款式表现上将西式套装与日本和服的腰带相结合。而在套装的局部，如宽大的袖形上也植入了和服元素，但选择了西式套装的经典剪裁，这些经典元素代表了东西文化共聚后的精致和高雅。在腰部加入了日本和服的腰带元素，既是一件实用的服装配饰，强化了西式服装中对女性身形的塑造方式，又是对日本传统服饰的致敬。腰带的材质和色彩也充分体现了日本文化的精美和细腻。在面料选择上，纹样选取了传统日本的扇形图案及花卉纹样等，以刺绣和印花相结合，既反映了西方时装的质感，也展示了东方面料的华丽。我认为设计应该代表一种文化的交流和融合，呈现出东西方文化的奇妙交汇点。这件作品不仅是时尚的，更是文化的新奇表达，以展现当今的时尚、艺术等文化的多样性与美感。

融合
纺织面料

Lan Xing 蓝星

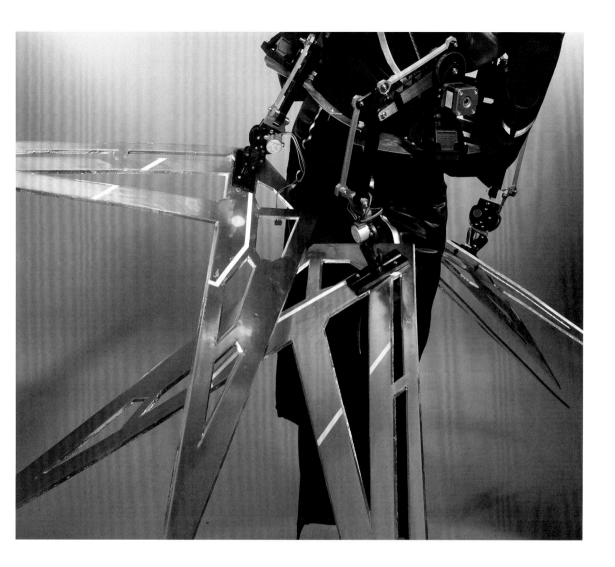

National patterns contain excellent national spirit and national culture, which deserve to be inherited. I have been thinking about what kind of graphic pattern will inherit and develop in the future? So, I take the sun pattern of the Dong people as the focus of this design.

The transformation from 2D pattern to 3D form, from static clothing to dynamic clothing evolution, from conventional fabrics to metal, as well as the interaction between people and clothing, breaking through the "tradition". By breaking the boundaries of traditional clothing, the work create a fascinating future visual effect and interactive experience.

To achieve the rotating, interactive induction of the future type of clothing, let more experiencers "participate" in it and feel the mystery of the Dong People's traditional sun pattern.

Future Graphicism
Leather, metal, 3D printing

民族纹样蕴含着优秀的民族精神、民族文化，它们值得被传承。我也一直在思考，未来图形纹样将会以什么样的形态传承与发展？于是我以侗族太阳纹为本次设计的重点。

该作品从 2D 纹样到 3D 形态的转化，从静态服装到动态服装的演变，从常规面料到金属，以及人与服装的交互行为等方面切入，突破"传统"。通过打破传统服装的各种常规思维界限，该作品营造出一种令人遐想的未来视觉效果和交互体验。

实现可转动、可交互感应张合的未来型服装，让更多的体验者"参与"其中，感受侗族传统太阳纹样的奥秘。

未来图形主义
皮革、金属、3D 打印

Louda Larrain (USA) 劳达·拉瑞恩（美国）

Through experience, exploration and experimenting Louda Larrain developed her own technique of fabric construction. It is her personal and unique way of creating an artwork with textile elements.
From a very young age Louda was involved in the art world. Drawing and painting were her passions and it is transparent in her textile work. She approaches her fabrics like tableaux that became alive in this video.

通过经验积累和不断的探索试验，劳达·拉瑞恩开发了个人专属的织物构建技术。这也是她用纺织元素进行艺术创作的独特方式。
劳达自幼就沉浸在艺术世界中。她对绘画的热爱在她的织物作品中体现得淋漓尽致。这段视频生动地展现了她对待织物就如同对待静物画一般的创作路径。

Painting with Fiber
Virtual works/Video format

纤维作画
虚拟作品 / 视频形式

Lee Yeonhee (Korea) 李莲姬(韩国)

It is a knit one-piece dress design using a Fair-Isle pattern, a traditional Nordic knitwear, and the pattern unique to the Fair-Isle pattern is changed by using the front and reverse. It has a voluminous silhouette with relaxed design.

这是一款使用传统北欧针织品费尔岛图案的针织连衣裙设计，通过使用正面和反面改变了费尔岛的独特图案，具有宽松的设计和丰满的轮廓。

Finding Fair-Isle
Wool, yarn hand knitting

寻找费尔岛
羊毛线、手工针织

Li Wei 李薇

Through 3D printing, the work realizes the two-dimensional and three-dimensional fusion between clothes and people, and the layering of shapes and blocks, the figurative and substantial clothes and the abstract and illusory shadows present an virtual and real space, which not only expresses the flexible artistic conception of the work, but also expresses the designer's personality and emotion.

作品通过 3D 打印，实现衣与人之间由二维及三维的交融，由形状到体块的层层叠叠，具象实质的衣和抽象虚幻的影呈现虚实相交的空间，既表达出作品灵动的意境，又表现了设计师的人格风度及个性情感。

Empty and Shadow
3D printing photosensitive resins

空与影
3D 打印光敏树脂

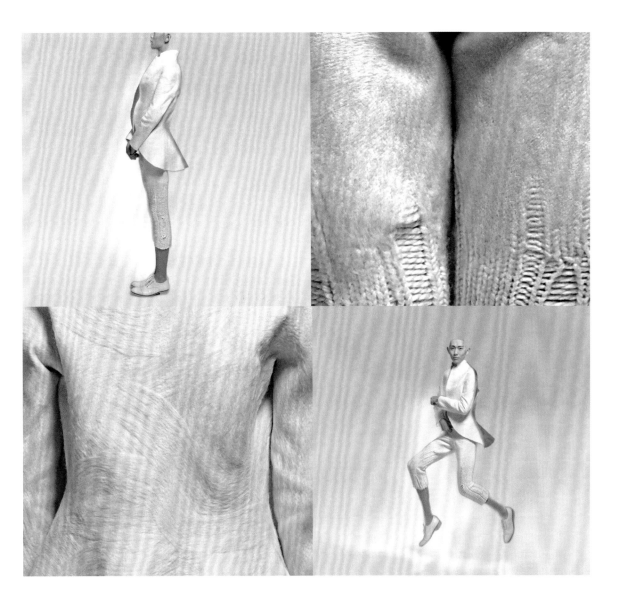

The work focus on the simple emotions and sparkling wisdom of the folk.

该作品关注民间质朴的情感与闪光的智慧。

Seamless Felt Clothing
Wool, cotton

毡衣无缝
羊毛、棉

Leonid Krykhtin (Britain/Russia) 利昂·克雷赫廷（英国/俄罗斯）

The thought-provoking video art film *Who Holds the Future* explores the intersection of AI, generative technology, and the future of human creativity. Specifically, it examines how advancements in AI will transform the fields of fashion design and art in the years to come. As algorithms become increasingly adept at generating original clothing designs, and even entire artworks, what role will human designers and artists play? Will they become obsolete, or will a new creative harmony emerge?
The film poses these pressing questions through an abstract narrative portrayed via avant-garde fashion visuals. The "AI" creates fantastical, impossible garments and artworks, far surpassing human imagination. The narrative suggests this is not the end - rather a transformation.The film's visual pace intensifies as the AI's fashion creations begin integrating with the abstract 3D art worlds. There is a sense of conflict and harmony at the same time as we start collaborating in the creative process.
The film starts and ends with an AI "entity"- portraying a future where humans embrace and partner with generative technology to push creative boundaries together. The core message is that AI does not replace human creativity - it evolves it. We remain integral creators, now elevated by imaginative new partnerships with technology.

《谁掌握未来》是一部发人深省的视频艺术电影，它探讨了人工智能、生成技术和人类创造力未来的交叉点，特别是人工智能的进步在未来几年将如何改变时装设计和艺术领域。随着算法越来越擅长生成原创服装设计，甚至整个艺术作品，人类设计师和艺术家将扮演什么角色？他们会被淘汰，还是会出现一种新的创作和谐？
影片通过前卫的时尚视觉效果，以抽象的叙事方式提出了这些紧迫的问题。人工智能创造出奇幻、不可能存在的服装和艺术品，远远超越了人类的想象力。叙事表明，这并不是终点，而是一种转变。随着人工智能的时尚创作开始与抽象的 3D 艺术世界融合，影片的视觉节奏也随之加快，此时在创作过程中的合作有一种冲突与和谐并存的感觉。
影片以一个名为"独立实体"的人工智能为开头和结尾，描绘了一个人类拥抱生成技术并与之合作，共同推动创造极限的未来。影片传达的核心信息是，人工智能不会取代人类的创造力，而是会促使人类的创造力不断进化。我们仍然是不可或缺的创造者。现在我们的创造力因为科技这一富有想象力的新助力得到进一步提升。

Who Hold the Future
Virtual works/Video format

谁掌握未来
虚拟作品 / 视频形式

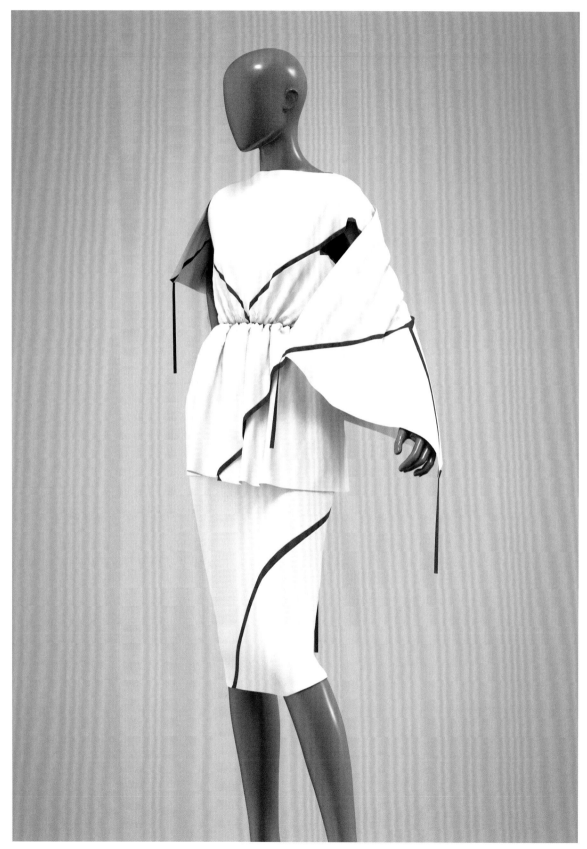

Liang Li 梁莉

Within a square, there can be many variations. Based on the concept of zero-waste cutting, various forms of laser cutting are carried out on the inside of square fabrics to form flat geometric plates of different shapes and sizes, which can be sewn together to form garments of different silhouettes.
A square inch is also the heart. There is a world within a square inch, and the human heart is rich and deep, so one can look into the heart to cope with the uncertainty of the outside world.

方寸之间,可以有诸多变化。以零浪费裁剪概念,在正方形布料内部进行各种形式的激光切割,形成形状各异、大小不一的平面几何板片,可以缝合为不同廓型的服装。方寸亦即内心。方寸之间自有天地,人心丰富而深邃,向内心观照,以应对外界的不确定性。

Within a Square
Cotton Fabric

方寸之间
棉布

The work explains that in the Chinese aesthetic field, white is colorless and multi-colored, and it is the most stunning color. The work is permeated with the philosophy of Lao Zi and Zhuang Zi to see the reality with emptiness and the aesthetics of knowing white and knowing black, the work has a graceful and elegant temperament, and collects the world's beautiful colors in the aesthetics of the body of the white body. The works have deep Chinese cultural literacy and the visual tension of modern design.

作品阐释了在中国的美学场域，白色即无色，也即多色，是绝艳的色彩。作品渗透老子、庄子以虚见实的哲学和知白识黑的美学境界，作品具有雍沉素雅的气度，敛天下秀色于虚白之躯的美学境界，具有深厚的中国文化素养和现代设计的视觉张力。

Colorless Stunning
Cashmere wool, jacquard cotton, wire organza, 3D hollow space cotton

绝艳无色
开司米毛线、提花棉布、金属丝欧根纱、3D镂空太空棉

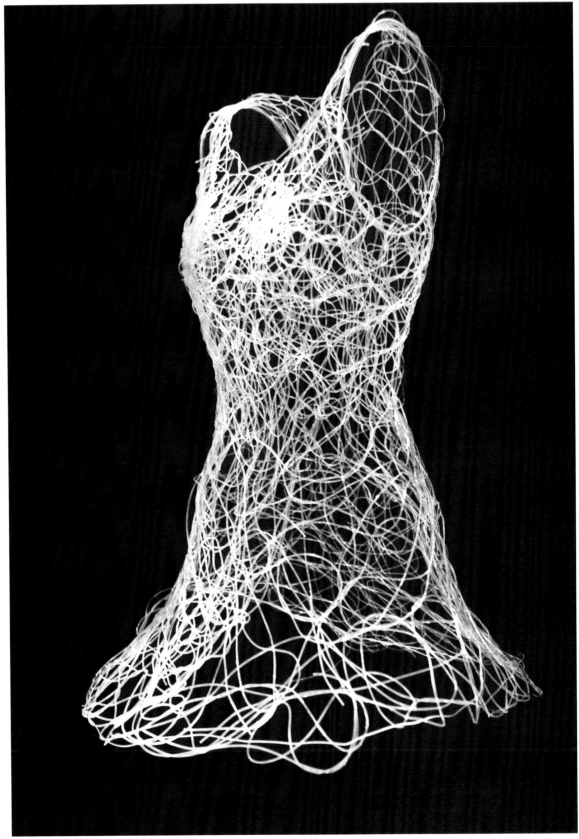

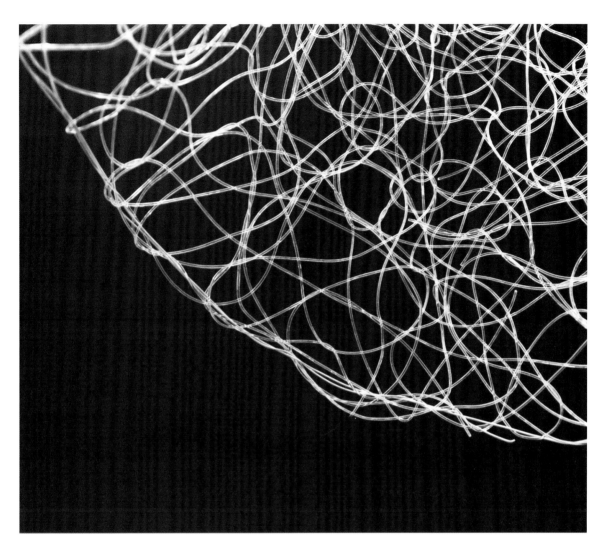

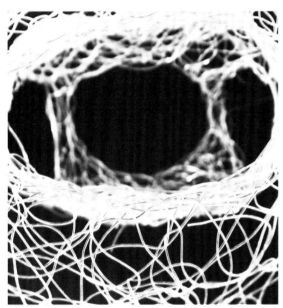

New Dress weaves clothes with Optical fibers. It takes clothing as the carrier, knitting as the process, and fiber as the material to combines art and technology, tradition and modernity together. The artwork focuses on the complex urban environment, to expounding the modern people's examination, contradiction, confusion, and struggle of their own spirit, and reappears the eager appeal of individual expression.

《新衣》用光导纤维织成衣物，以服饰为载体，编织为工艺，纤维为材质，将艺术与科技、传统与现代相结合，阐述了在错综复杂的环境下，现代人对内心的审视、矛盾、困惑与挣扎，再现个体表达的迫切诉求。

New Dress
Optical Fibers

新衣
光导纤维

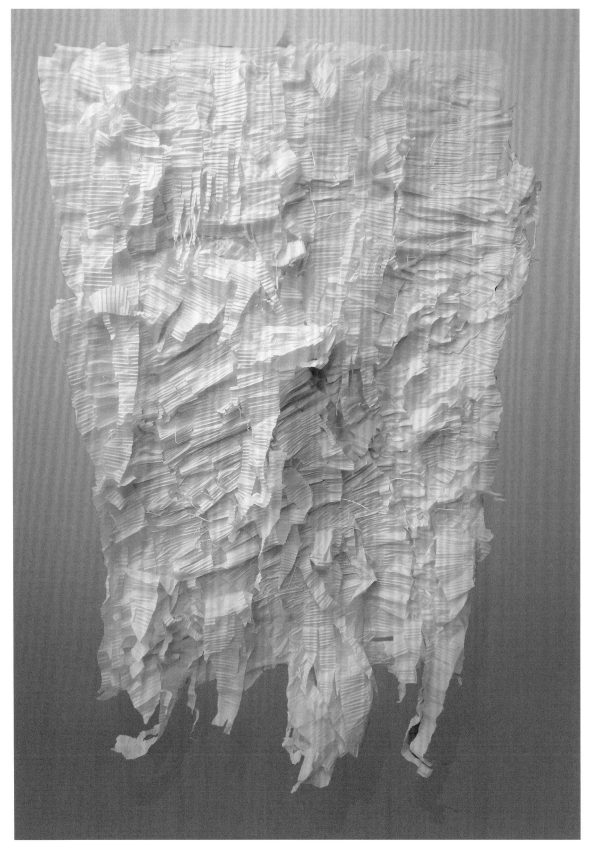

Liu Hui 刘辉

Another Life created from common paper materials, which are folded, stuck, torn, and twisted to form new visual and tactile forms. The layers intertwine and embrace each other, creating a transformation of time and space that echoes the seemingly simple repetition of daily life. Despite its simplicity, this art piece embodies a resilient and powerful force within everyday life.

《另一种生活》材料为日常生活中常见的纸，通过压褶、粘贴、撕扯、捻揉，构成了不为人们所熟悉的视觉与触觉形态，层叠中穿插、彼此互拥。这种融合了时间与空间的变化呼应着一种蕴含无限变量却看似简单、重复的日常生活，平凡无奇却孕育着一种坚韧而强大的力量。

Another Life
Cotton, paper

另一种生活
棉、纸

Liu Jing/Kinor Jiang (Hongkong, China) 刘静 / 姜绶祥（中国香港）

Coating polyester with titanium metal on the surface is a modern and sustainable production technique that imparts vibrant luster to traditional fabrics. A strap, like a link of time, bridging traditional garments with modern technology. The incorporation of metallic fabric bestows a distinctive texture upon the entire garment, emitting a futuristic sheen. Simultaneously, this clothing piece amalgamates traditional Chinese tailoring with modern streamlined designs, resulting in a cutting-edge yet elegantly poetic visual effect. From different angles, the metallic sheen and the sweeping lines of the robe harmonize, accentuating the graceful movement of the waistband. The overall silhouette flows seamlessly, exhibiting a unique rhythm of beauty with each step taken.

在涤纶表面镀覆钛金属，是一种能够赋予传统布料鲜艳光泽的现代可持续生产技术。而一条系带，仿佛是时光的纽带，将传统服装与现代科技连接在了一起。金属布的运用赋予了整件服装独特的质感，使其散发着未来感般的光泽。与此同时，该服装作品将传统的中式剪裁与现代的流线造型相结合，塑造出前卫而又充满典雅诗意的视觉效果。不同角度下，金属光泽与荡领的线条交相呼应，映衬着腰间灵动的系带，整体廓型流畅，在走动间展现出独特的韵律美。

Tied 1
Polyester, titanium

结 1
涤纶、钛

Liu Qin 刘沁

The *Wormhole* series of works is a combination of inflatable technology in the context of space exploration, to develop some associations with wormholes.
Einstein proposed the wormhole theory, which simply means "wormholes" are spatiotemporal tubes connecting distant regions of the universe. Wormholes are the intersection of time and space in the universe, and the legend that crossing through wormholes can lead to parallel worlds has inspired all scientists from ancient times to present, and they are willing to continuously explore them.
Inspired by the exploration and contemplation of unknown parallel spaces, the *Wormhole* series designs immerse human fantasies about the universe in clothing language and endow clothing with new ideas through new technologies.

《虫洞》系列作品是在宇宙探索的背景下，结合充气技术，展开对虫洞的一些联想。
爱因斯坦提出虫洞理论，简单来说，"虫洞"就是连接宇宙遥远区域的时空细管。虫洞是宇宙中时空的交错点，而穿越虫洞可以抵达平行世界的传说，让从古至今所有的科学家都心生向往，并愿意为之不断探索。
有感于对未知平行空间的探索与思考，《虫洞》系列设计将人类对宇宙的幻想用服装语言倾情创作，以新技术赋予服装新思路。

Wormhole
Inflatable composite fabric, milk silk, fabric metal

虫洞
充气复合面料、牛奶丝面料、金属

Ocean, sunrise, glowing rays, boat oars, stars, the bright moon... Seeking color inspiration from the variegated nature, in the serene projection and natural tracing, blue, white, purple, orange and other colors overlap and grow, either in collision or in harmony, presenting a new symbiotic relationship between man and nature, tradition and modernity, art and humanities, while upholding rationality and sensibility together, healing and nourishing the new sustainable fashion.

海洋、日出、霞韵、船橹、星彩、皓月……从形色斑斓的自然万物中寻觅色彩灵感,在宁静的投映和天然的描摹中,蓝、白、紫、橙等各式色彩交叠丛生,或对撞或和谐的组成形态,呈现人与自然、传统与现代、艺术与人文间崭新的共生关系,同时秉持理性和感性共进的方式,疗愈和滋养崭新的可持续时尚。

To View the Boundless Ocean from the Eastern Shore
Fabric

东临碣石　以观沧海
纺织面料

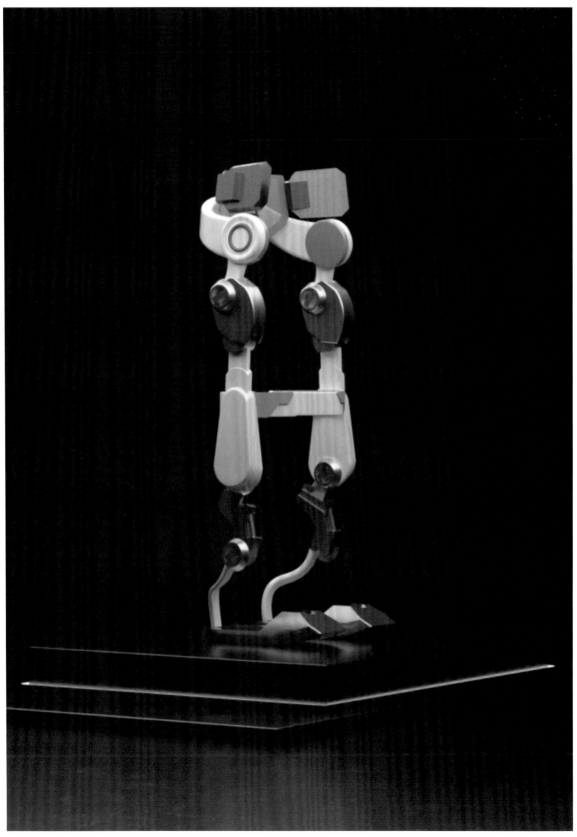

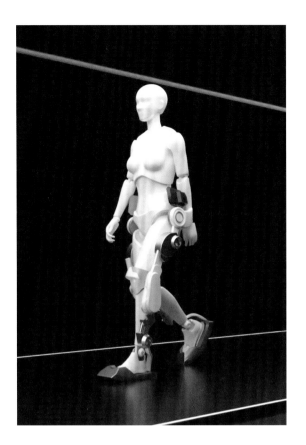 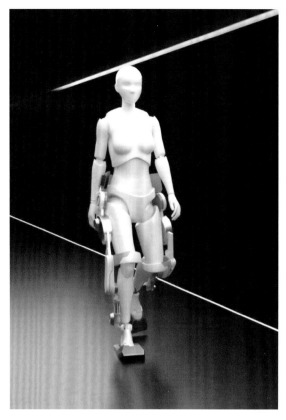

Redefinition of "Body": Human-Machine "Co-Integration" Exoskeleton is committed to starting from the perceptual three-level theory, triggering the psychological level of the "human" perspective by stimulating the physiological level of the product perspective of human-machine "integration", and from the "system" perspective, exploring the systematic relationship between human and machine integration, and constructing a system model of human-machine cooperation, resolve deep cultural conflicts between humans and machines. And through the design of compensatory methods, combined with new materials and technologies, we aim to improve the safety, usability, comfort, and interactive experience of human-machine "co-fusion" exoskeletons, making them more efficient, comfortable, convenient, safe, and operable, in order to reduce the "exclusivity" of extended products such as exoskeletons, thus achieving the integration of human and machine exoskeletons with users in terms of physical functionality, styling, and emotional integration in terms of psychology, spirit and emotions.

《"身体"再定义：人机"共融"外骨骼》致力于从感性三层次理论出发，通过对人机"共融"产品视角的生理层面刺激触发"人"视角的心理层面，并从"系统"视角出发将产品维度功效和经济性与情感维度情绪和体验度出发，探讨人与机器融合的系统关系，构建人一机协作的系统模型，解决人一机深层次文化冲突，并通过设计代偿的方法，结合新材料、新技术，提升人机"共融"外骨骼的安全性、易用性、可用性、使用舒适性、互动体验性等，让人机"共融"外骨骼更加高效、舒适、便捷、安全和可操作，以减少身体本体对外骨骼等外设延伸产品的"排异"性，从而实现人机"共融"外骨骼与用户在物理层面功能、造型及感性层面的心理、精神和情感上的共融。

Redefinition of "Body": Human-Machine "Co-Integration" Exoskeleton Design　"身体"再定义：人机"共融"外骨骼设计
Titanium Alloy, anodized alumina, carbon fiber, silicone fireproof cloth, electronic skin, etc.　钛合金、阳极氧化铝、碳纤维、硅胶防火布、电子皮肤等

Sanchia Lau (Macao, China) 刘欣珏（中国澳门）

Explore the real world inside, plant the seeds of hope, and start from the heart. A group of healing dolls travel around the world, bringing healing power to people! Behind any uncertain tomorrow, in fact, there is a certain present, hope and love, we might as well live a little more boldly. Let the doll's journey become our inner journey, embark on the journey, explore the heart!

探寻内心的真实世界,种下希望的种子,从心出发。一群带着疗愈能量的娃娃们穿越世界各地,不断给人们带来疗愈的力量!任何不确定的明天背后,其实都有着确定的当下、希望与爱,我们不妨活得再大胆一点。让娃娃们的旅行成为我们内心的寻找之旅,踏上旅程,探寻内心!

Chiwawa: Country Road of Healing Art
Virtual works/Video format

祈愿娃娃们的疗愈艺术之旅
虚拟作品 / 视频形式

Liu Xin 刘鑫

Ntopia originally referred to "Utopia" and attempted to construct this "anti utopian" world by reversing the first letter. At the time when Cyberspace was the ontology of post human research, the body became the "antenna" for the bidirectional construction of virtual and real worlds in terms of technology, art, fashion, and society. With the increasing attack of artificial intelligence on us, humanity's "expression power" in the virtual world is gradually diminishing, and the control of the virtual world's body and identity is gradually losing control. Humanity is gradually "forgetting" the unique body. The work uses digital modeling and AI assistance as the basic workflow, describing a future utopian world. As an allegory of art and criticism, the entanglement and ferocity of the body symbolize the chaos and visual frenzy of the virtual world. The camera, which has been alienated by industry, is an external "organ", and multiple sensory dimensions reflect the enhancement of the body in the human world.

Fashion, as an important material carrier of human aesthetics, is intertwined and intertwined with the body, implying the spiritual boundaries and fixed aesthetic paradigms that humans find difficult to break. Human beings no longer have physical bodies in virtual space, and the spirit transforms into desire machines. Virtual bodies become slaves to embodied intelligence and visual representation. From the perspective of fashion supervision, the physical enhancement of virtual bodies allows technology to extend body functions in both time and space, and also makes the body and spirit the subject of supervision. The beauty of the freely growing body is completely eliminated, replaced by a body that has been "disciplined".

"N" topia
Virtual works/Video format

Ntopia 原指"Utopia（乌托邦）"，通过首字母颠倒试图构建这个"反"乌托邦世界。在赛博格作为后人类研究的本体论之时，身体之于技术、艺术、时尚、社会则成为虚拟与现实世界双向构建的"天线"。

随着人工智能对我们的攻势越发强烈，人类在虚拟世界的"表达权"正逐渐式微，虚拟世界身体和身份的掌握正逐渐失控，人类正在逐渐"遗忘"唯一性的身体。作品以数字建模与AI辅助作为基本工作流，描述了一个未来的异托邦世界。作为艺术与批判性的寓言，身体的缠绕、狰狞象征着虚拟世界的群魔乱象与视觉狂热，被工业异化的摄像机是外置"器官"，多重感官维度映射身体在人类世界中的身体增强。

时装作为人类审美的重要物质载体，与身体缠绕与交织，暗示着人类难以打破的精神边界和固化的审美范式。人类在虚拟空间再无肉身，精神幻化欲望机器，虚拟身体成为具身智能与视觉表征的奴隶。从时尚监督的维度来说，虚拟身体具身性的身体增强，使得技术在时间和空间上延伸了身体的功能，也使身体精神成为被监督者。自由生长的身体之美荡然无存，取而代之的是被"规训"的身体。

反乌托邦
虚拟作品/视频形式

Inspired by the poem *Tao Ge* (song of the pottery) by Gong Shi of the Qing Dynasty -
"In the fire white glaze and blue flowers are made.
The flowers are clear from the glaze.
The innate wonder of creation can be seen.
The infinity is born from the Taiji."
A Thread & Blue and White, the infinite thread, from nothing to something expresses the coexistence of man and nature; man, heaven and earth, the sun and the moon correspond to each other.
My mind is silent and thoughtless, all goodness has not yet been developed, but there is a clear and unmistakable essence.

A Thread & Blue and White
Wool

创作灵感来源于清代龚轼的《陶歌》——
"白釉青花一火成，花从釉里透分明。
可参造化先天妙，无极由来太极生。"
《一根线 & 青花》，无限之线，从无到有表达人与自然共存；人与天地相参，与日月相应。
吾心寂然无思，万善未发，然有昭然不昧之本体。

一根线 & 青花
羊毛

YT.LIU (Australia) 刘伊童（澳大利亚）

The showcased dress and jacket ensemble from the "PULSE" series not only inherits the exquisite craftsmanship of high fashion, but also cleverly blends the comfort and vitality of sportswear. With an inner dress that fits lightly and snugly, like a layer of soft skin, juxtaposed with the outer down jacket, the shoulders are sculpted to resemble divine beings, full and towering, creating a perfect hourglass silhouette. Each interwoven structure is a boundless imagination of the future. Regardless of time or place, the wearer can feel the surge of vitality, as if their body is embraced by the wind, bravely stepping into the unknown. At the same time, the design is filled with futuristic elements, and every hollow and flowing line outlines a magical scene of unknown galaxies and streaming data, presenting a seamless integration of the wearer with technology.

This bold fashion endeavour is a positive response to diversity and an immersive experience. Wearers can freely roam on the streets, at events, or even in futuristic alternate dimensions, showcasing their confidence and unique beauty, seamlessly merging sportswear's energy with high fashion's elegance.

PULSE
Jersey, mesh, PU

《脉冲》系列的秀款连衣裙夹克套装不仅继承了高级时装的精湛工艺，更巧妙融汇了运动装的舒适与活力。
该作品内外搭配，连衣裙轻盈贴合，仿佛是一层柔软的肌肤，和外层羽绒夹克形成鲜明对比，肩部塑造得宛若天神，饱满、高耸，令整体线条呈现完美的沙漏曲线。每一处交织的构造都仿佛是对未来的无限遐想。无论何时何地，穿着者都能感受到活力的涌动，仿佛身躯融入风的怀抱，勇敢地迈向未知。与此同时，设计中充溢着未来感的元素，每一处镂空和流线勾勒出未知星系与流动数据的幻景，将穿着者与科技的融合呈现得淋漓尽致。这种时尚的大胆尝试，不仅是对多元化的积极回应，更是一种身临其境的体验。穿着者无论在街头、宴会，还是未来异次元空间中，皆能自由驰骋，展示出自信与独特之美，将运动的活力与高级时装的优雅融为一体，创造出全新的时尚美学。

脉冲
弹力网纱、弹力针织、环保皮革

YT.LIU (Australia)　　　　　　　　　　刘伊童（澳大利亚）

MQUEYT's digital garments coexist with immense potential in physical reality, capable of evolving into genuine fashion masterpieces. We aspire to ensure that each design showcases robust functionality, captivating interactivity, and vivid dynamism within the YTOPIA METAVERSE. These designs merge two layers of material, one tightly adhering to the body and the other featuring vast glass-like contours. This combination perfectly showcases MQUEYT's daring and iconic style in the virtual world, featuring seamless cuts, broad shoulders, cinched waists, and a unique V-shaped silhouette. The glass-like sections on the digital garments adapt to the surrounding environment and the curves of the human body and engage in mutual resonance. These designs are not only dynamic but also highly interactive while still retaining practical usability.

MQUEYT 的数字服装在现实物理空间中与巨大的潜力共存，可以进一步演化为真实的时尚佳作。我们追求每一件设计都能在乌托邦元宇宙中展现出强大的功能性、引人入胜的互动性和生动的动态性。这些设计融合了两层材料，一层紧密贴于身体，另一层则采用巨大的玻璃状廓型，完美呈现出 MQUEYT 在虚拟世界中大胆的标志性风格，诸如流畅的剪裁，宽阔的肩部，收腰和独特的 V 形腰线。服装上玻璃般的部分不仅能够自适应周围环境和人体曲线，还能与之相互呼应。这些设计不仅富有动感，而且极具互动性，同时又具备实际的使用功能。

Lucid Flux
Virtual works/Video format

澄澈流动
虚拟作品 / 视频形式

Worsted cotton cloth is cut into diagonal strips to make 2.5mm cloth rope, and then the structure and form of the human body one by one to produce a modern form of traditional style clothing, the use of sparse and dense combination of lines to achieve which is static and dynamic, dense and sparse, traditional and modern.

把精纺棉布裁成斜条做成粗 2.5mm 布绳，然后以人体的结构、形态，一条一条地制作出形式现代的传统风格的服饰，利用线的疏密组合，达到既静又动、既密又疏、既传统又现代的效果。

The Flow and Restraint of the Line
Worsted cotton

线的流动与克制
精纺棉布

The creative inspiration for this series of digital prints is derived from the digital weaving transformation of traditional ethnic minority embroidery pieces. Analyze and record the collected embroidery pieces in terms of images, text, craftsmanship, and form a basic database through digital archive collection. By utilizing AI technology and integrating virtual design, innovative simulation design of embroidery pieces is carried out through storage processing, model creation, virtual simulation, restoration and reproduction.

This avant-garde illustration draws inspiration from the classical verse "Budding faint green, touched by a hint of spring," seamlessly blending traditional embroidery motifs with contemporary landscape elements. The overarching color palette centers around a gentle shade of green, symbolizing the tender emergence of early spring. This delicate hue will be subtly woven into the embroidery elements, outlining the contours of flowers and leaves, infusing the illustration with vitality. Concurrently, modern mountain and water features will be portrayed in deep shades of gray, creating a striking contrast with the green and evoking a sense of the passage of time.

A Hint of Spring
Virtual works/Video format

该系列数字版画的创作灵感源自传统少数民族绣片的数字织造转化。对采集的绣片进行图像、文字、工艺等环节解析记录，以数字化档案采集形成基础数据库。运用AI技术，结合虚拟设计整合，以存储处理、模型创建、虚拟仿真、复原再现等方式进行绣片的创新模拟设计。该作品在创作中围绕"古绣新生"之意，以数据筛选出"才吐微茫绿，初沾少许春"两句古诗进行时尚创意，旨在将传统的绣片艺术与现代时尚美学相结合，又将现代性的极简山水元素相融合，营造出一幅充满诗意与现代感的画面。整体色彩选取以柔和的绿色为主基调，象征着早春初放的嫩绿。这一绿色也在绣片的元素中得以体现。山水、花叶的轮廓以淡绿线条勾勒，为整幅插画增添生机。同时，通过深邃的金属质感灰色，来与古典绣片服饰纹样形成碰撞与鲜明的对比，传递出时光流转的感觉。

少许春
虚拟作品 / 视频形式

This work is a cross disciplinary practice in various fields such as digital weaving technology, classic gold repair art, and fashion clothing art, providing modern digital weaving technology with more ethnic cultural connotations to shape the image and achieve traditional fashion transformation. Based on cross-border thinking, new creativity is infused into traditional culture, enabling innovation and promotion of national and folk aesthetics.

Especially on the basis of refining the elements of ethnic embroidery, new technologies and expression methods have been used to express clothing with strong ethnic characteristics and unique aesthetic styles, in order to showcase the unique cultural and artistic charm of the region. The exhibited works *Kylin* are taken from one pieces in the series *New Variables*.

该作品以数字织造技术、经典金缮艺术、时尚服饰艺术等多领域跨界实践，为现代的数字织造技术提供更加具备民族文化内涵的形象塑造，实现传统的时尚转化。以跨界思维为基础将新的创意注入传统文化中，让民族和民间美学得以创新张扬。

尤其在提炼民族刺绣元素的基础上，该作品以新的技术和表现手段呈现了具有浓厚民族特色和独特美学风格的着装，以展现该地区独特的文化和艺术魅力。此次展出作品《麒麟》取自系列作品《新变量》中的一件。

New Variable—Kylin
3D printing PLA, recycling ethnic embroidery pieces, composite silk satin chiffon, etc

新变量——麒麟
高韧性 3D 打印、回收少数民族绣片、印花素绉缎等

A total of 14 key ancient buildings were damaged in the powerful 8.1 magnitude earthquake in Nepal in 2015; Notre Dame de Paris suffered a fire in 2019, which severely damaged the overall building, with most of the top burned and the spire collapsing in the fire.

Earthquakes, fires and wars are all destroying and damaging human cultural heritage and monuments. With the theme of remodeling, the work uses the sense of conflict formed by the upside-down and broken frame and the delicate and exquisite handwork, and the contrast formed by the coarse wool felt and the silky and light fabric to express the feelings of passing away, remembrance and deploring.

2015年尼泊尔8.1级强震，共计损毁了14座重点古建筑；2019年巴黎圣母院遭遇大火，整体建筑损毁严重，大部分顶部被烧毁，塔尖在大火中倒塌。

地震、火灾、战争无一不在摧毁和破坏人类的文化遗产和古迹。作品以重塑为题，用颠倒破损的框架和精致细腻的手工形成冲突感，用粗砺的羊毛毡和丝滑轻薄的面料形成对比，表达逝去、追忆、痛惜的感受。

Reminisce
Plant-dyed Wool felt, silk, cotton and linen

追忆
植物染的羊毛毡、丝、麻

Lyu Yue (Aluna)　　　吕越

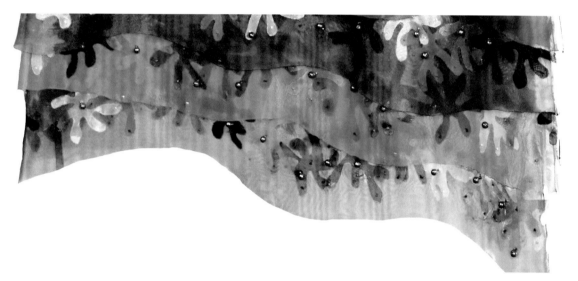

The works named *Depth of Flowers* is composed of transparent yarn feeding a group of cold colors and colorful metal sheets engraved with Chinese characters "花". The plane tailoring makes the works seem to have some Oriental charm. And the transparent material czexpresses a kind of relaxed and happy effect.

《百花深处》是用一组冷色调的透明纱料加上激光雕刻汉字"花"的彩色金属片制成的作品。平面的剪裁结构使作品看上去有一些东方韵味,透明的材料想要表达一种轻松愉快的效果。

Depth of Flowers
Organdie, organdy silk

百花深处
蝉翼纱、丝绸

Melissa Coleman/Leonie Smelt (Netherlands)

Every 11 minutes a woman dies from an unsafe abortion. This dress shivers with light in sympathy for each one.
The dress is inspired by our ability to measure the seismic waves produced by an earthquake all across the planet and wishes that the loss of life could do the same. It translates a raw number from a Unicef report into a timed event that sends a message to other women. It emphasizes our interconnectedness as people, using technology to create a poetic event that makes the wearer aware of the loss of an unknown life and mourn it.

每 11 分钟就有一名妇女死于不安全堕胎。这条裙子散发出的光芒让人不寒而栗，对每一名死去的妇女表示同情。
这条裙子的灵感来自我们测量地球上地震产生的地震波的能力，并希望失去的生命也能产生同样的地震波。它将联合国儿童基金会报告中的一个原始数字转化为一个定时事件，向其他女性传递信息。它强调了人与人之间的相互联系，利用技术创造了一个诗意的事件，让穿戴者意识到一个未知生命的逝去并为之哀悼。

Tremor
Fabric, paper, electronics with LEDs

震颤
织物、纸张、LED 电子设备

Fawad Noori/Suwaiba Fawad (Pakistan) 法瓦德·努里 / 苏维巴·法瓦德（巴基斯坦）

"Epoch Diva" is a Rhythm of Mix culture through Fashion. This couture outfit is the inspiration of bliss full natural beauty of ancient architecture of Puyuan Town that is one of the "Five Famous Towns" in Jiangnan area during the Ming and Qing Dynasties. It is eminent for its well-preserved ancient architecture and traditional Chinese culture. The personalized digital printed Silk fabric design with the combination of scenic view of House, lakes and structure of ancient architecture with illustrious blue pottery of Pakistani culture locally known as kashi work in Multan developed its own unique, indigenous style. Blue pottery enjoyed patronage of the royalty and used as utensils, decorative tiles and architecture. The outfit Epoch Diva naturally, spiritually and romantically balanced couture outfit with a royal and luxury touch. The 3D embellishment enhance its combination of cultural beauty as if she is actually wearing a spring garden. The skill for creating this blue pottery, also known as kashi work, was introduce centuries ago by local artisans, whose craft derived influences from Persia, Central Asia and the Mongols. It is widely believed that kashi works originates from Kashgar, a city in western China.

Epoch Diva: Rhythm of Mix culture through Fashion
Silk, cotton, patchwork, digital printed Self design fabric, crushed pleated fabric, trims and Finishes Embellishment material with hand embroidery

《纪元天后》通过时装展现多元文化的律动。这套高级定制服装的灵感来源于明清时期江南"五大名镇"之一的濮院古镇中充满幸福感的自然美景。濮院镇是明清时期江南"五大名镇"之一，以其保存完好的古建筑和中国传统文化闻名于世。

个性化的数码印花丝绸面料设计将木尔坦的房屋、湖泊和古建筑结构等美景与巴基斯坦当地著名的喀什作品相结合，形成了自己独特的本土风格。蓝陶受到皇室的青睐，被用作器皿、装饰瓷砖和建筑材料。《纪元天后》的服装在自然、精神和浪漫之间取得了平衡，具有皇家和奢华的气息。3D点缀增强了其文化美感，仿佛真的身着春色满园。几百年前，当地工匠就开始制作这种蓝陶，也被称为"喀什作品"，其工艺受到波斯、中亚和蒙古人的影响。人们普遍认为，喀什作品起源于中国西部城市喀什。

纪元天后：多元文化协同时尚律动
丝绸、棉布、拼布、数码印花自我设计织物、碎褶织物、饰边和饰面、手工刺绣装饰材料

Mu Yun 穆芸

The theme of the exhibition work *"To weave with the legendary landscape"* is centered around the famous town of the Yangtze River with a history of nearly one thousand years. It not only has the beautiful scenery of the Yangtze River, but also has a long tradition of textile industry, and has the reputation of "the first city of sweaters in China" and "the hometown of woolen sweaters". Through the traditional knitting machine, organized with different texture technology and the modeling design of fashion art, the works represent the beautiful cultural landscape of Puyuan town and the clear water and green mountains of the vibrant characteristic wool knitting industry.

参展作品《织衫织水》，主题围绕着有着近千年悠久历史的江南名镇濮院，它不仅有江南清丽的山水风光，更有悠久的纺织业织造传统，具有"中国毛衫第一市"和"羊毛衫之乡"的美誉。作品通过传统的针织机编，以不同肌理工艺组织，以时装艺术的造型设计，表现出代表濮院镇秀丽多姿的人文景观和生机勃勃的特色毛针织产业。

To weave with the legendary landscape
Knitting machine

织衫织水
针织机编

The main color is black and white with simple style and soft lines. She chooses flowing line patterns to express the rhythmic beauty of Chinese mountains and rivers, thus creating a pure and elegant visual effect in a simple and abundant artistic sense. By expressing the flow and change of mountains and rivers, as well as the ethereal and balance, she brings the harmonious coexistence with nature in poetic and serene clothing styles.

服装以黑白色为主调，款式以简洁、柔和的线条为主，选用流动的线条图案表现中国山水的律动之美，创造出一种纯粹而典雅的视觉效果，旨在呈现简约而又富有内涵的艺术感。通过表现山水的流动与变化，以及山水间的空灵和平衡，带来与自然和谐相处的感受，创造一种充满诗意和宁静的服装风格。

Void Landscape
Sweater coil, silk gauze

山水空灵
毛衣线圈、真丝绡

Johanna Braitbart (France) 乔安娜·布拉特巴特（法国）

Inspired by the colors of Puyuan Ancient Town, and the savoir-faire of this town, the *Forest Angel* is made of wool, feathers and beads. This international creation is made of European and Chinese culture.

《森林天使》的创作灵感来源于濮院古镇的风景色彩和文化精髓,由羊毛、羽毛和钉珠塑造而成。这是一件融合欧洲和中国文化的国际性作品。

Forest Angel
Feathers, beads, lace, silk, wool

森林天使
羽毛、玻璃钉珠、蕾丝、丝绸、羊毛

Qiao Dan 乔丹

The story that the earth with a Type5 civilization has gone through wars and epidemics, unrest and turmoil in 2023. It travels through the multiverse and begins to explore the higher-level civilizations that must exist in the universe. It opens the SIVICO Space base and can master the creation energy. The sixth-level civilization can manipulate time and space arbitrarily. When the earth was affected by 2019-nCoV, it opened the 61st second protection for it. After 1000000000 (billion) seconds, it helped the earth restore peace…
SIVICO 23S/S collection have "practical natural healing", "quiet lightweight protection", "functional tactile awakening-ignited aesthetics". The 15 looks on display will encapsulate the three waves of this SS23 collection.
The first wave theme is the 61st second protection, the style is practical natural healing and sedate lightweight protection, casual light business will include some practical lightweight protection details.
The second wave of theme formultidimensional space, the style is practical natural healing and functional tactile awakening, will be in the functionality of the expansion of the city, displaying leisure, functional fusion.
The third wave's theme is Type 7, existence or nonexistence, styled in a combustible out-of-this-world aesthetic that favours structural creativity and will have a breakthrough in the visual plasticity of the jackets.

本作品讲述五级文明的地球在 2023 年经过战争和疫情，不安和动荡，穿越多元宇宙，开始探索宇宙中一定存在的等级更高的文明，开启了斯威克时空基地，可以掌握创世能量的六级文明，可以任意操控时间和空间，在地球受到 2019-nCoV 影响时，开启了对它的第 61 秒保护，在 1000000000（十亿）秒后，帮助地球恢复了平静……
总体的设计风格为"实用的自然治愈""沉静的轻量防护""机能的触觉唤醒"与"燃起的出世美学"，现展示的 15 个画面将概括本次 23 春夏系列的三个波段。
第一波段主题为第 61 秒保护，风格是实用的自然治愈与沉静的轻量防护，休闲轻商务中会加入一些实用的轻量防护细节。
第二波段主题为多维空间，风格是实用的自然治愈与机能的触觉唤醒，会在功能性的城市拓展中，展现休闲、机能的融合。

第三波段主题为 7 级文明（存在或不存在），风格是燃起的出世美学，偏向结构的创意，将在夹克的视觉塑形中有所突破。

SIVICO Space base
Virtual works/Video format

斯威克时空基地
虚拟作品 / 视频形式

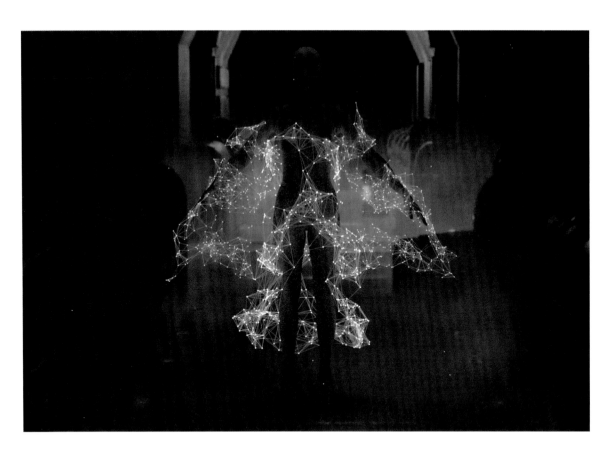

Qin Geng 秦耕

This work utilizes mapping technology to project virtual clothing onto the wearer, achieving a physical interaction between the virtual and the real world. Through this method, the audience can intuitively experience the aesthetic beauty of the virtual clothing and gain a deeper understanding of the wearer's physical characteristics. At the same time, the wearer becomes an indispensable part of the work, with their physical features complementing and echoing with the virtual clothing, forming a new form of bodily language.

Digital Equivalents—a virtual mapping of the reality clothing
Virtual works/Video format

此作品运用了 mapping 技术，将虚拟服饰投射在穿戴者身上，实现了虚拟与现实的身体交互。通过这种方式，观众可以直观地感受到虚拟服装的交互美感，并深入了解穿戴者的身体特征。与此同时，穿戴者成为作品中不可或缺的一部分，他们的身体特征与虚拟服装相互映衬、相互呼应，形成了一种全新的身体语言。

数字等身——虚拟服饰的现实映射
虚拟作品 / 视频形式

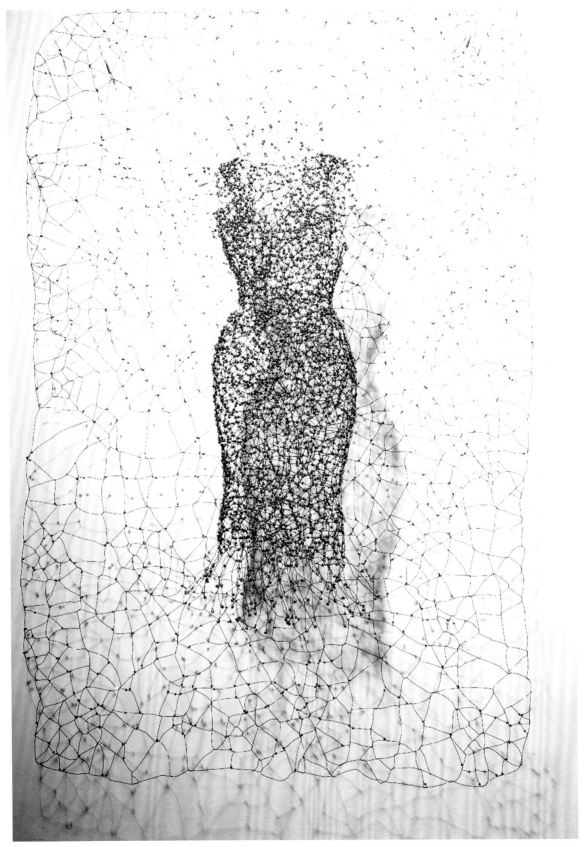

Key-Sook Geum (Korea)

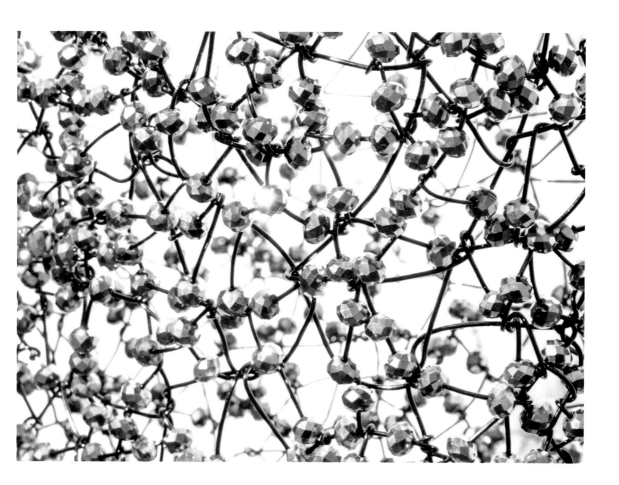

I tried to interpret the beauty of traditional architecture which is found in old cities as fashion art. Numerous dots made of roof tiles and flexible curves that touch the sky reveal the overall shape beautifully. Beads that I love play the role of dots, and numerous wires express curves and furthermore, become elements in constructing forms.
Encountering the natural connection between tradition and Fashion art, I feel excited and moved.

我试图用时尚艺术来诠释古城传统建筑之美。由瓦片组成的无数圆点和触及天空的柔性曲线将整体造型展现得淋漓尽致。我喜欢的珠子扮演了点的角色，而无数的线则表现了曲线，并进一步成为构建形式的元素。
传统与时尚艺术之间的自然联系让我感到兴奋和感动。

Dots to Line/Shapes to Form
Wire, beads

由点成线 / 由面到体
金属丝、珠子

The *Magnetic Response* project was initially inspired by one of my friends majoring in Physics. Their lab's magnetic fluid music demonstration device, which perfectly combined the magnetic liquid with the music art, caught my attention. After investigation, the motion curves of magnetofluid became my primary design materials. I made the magnetofluid cut the magnetic field, which I created using Rhino. The obtained linear structure offered inspirations to my garments and textiles.

This project, wrapped in a magnetic field, with music as a boat and clothing as another medium involved, sends out an answer on the other shore.

《磁应》项目的前期灵感来源于我学习物理专业的朋友分享给我的磁流体音乐演示装置实验视频。实验中磁流体磁控相变的物理特性和声音被转化为电再转化为磁流体动能的过程让我产生了强烈兴趣。经过调研学习，我开始思考如何通过服装的设计语言来诉说磁流体将声音可视化的这一过程。后期我通过建模软件对磁流体的动势及声波对应的变化进行了分析，得到了新的线性关系和立体结构，这也成为最后我服装造型和面料的主要设计支撑。

这个项目在磁场的包裹中，以音乐为舟，服装作为另一种介质参与其中，在另一岸发出了应答。

Magnetic Response
Virtual works/Video format

磁应
虚拟作品 / 视频形式

Sarah Siewert (Poland/Germany)

In a world surrounded by residential and commercial waste, my role as fashion designers is clear: In order to inspire innovations in textile recycling, I channel our emotions to discern the natural elements and use them to create new sculptural practices.

In my years of working with recycled and upcycled materials, I have drawn from a wide range of cultural, historical, and intellectual approaches to shine a light on the hidden bridge between natural aesthetics and the functional world of industrialized mankind.

在这个被住宅和商业废弃物包围的世界里，作为时装设计师，我的角色很明确：为了纺织品循环的创新，我引导自身情感去辨别自然元素，并用它们来创造新的雕塑实践。

我拥有多年使用回收和升级再造材料的工作经验，从广泛的文化、历史和思想路径中汲取灵感，在自然美学与工业化人类的功能世界之间架起了一座隐秘的桥梁。

The Appearance of a Circle
100% organza silk、Fishbone crinoline、Aluminium foil

圆之貌
100% 欧根纱真丝、鱼骨撑、铝箔纸

Sarah Siewert (Poland/Germany)

In a world surrounded by residential and commercial waste, my role as fashion designers is clear: In order to inspire innovations in textile recycling, I channel our emotions to discern the natural elements and use them to create new sculptural practices.
In my years of working with recycled and upcycled materials, I have drawn from a wide range of cultural, historical, and intellectual approaches to shine a light on the hidden bridge between natural aesthetics and the functional world of industrialized mankind.

在这个被住宅和商业废弃物包围的世界里，作为时装设计师，我的角色很明确：为了纺织品循环的创新，我引导自身情感去辨别自然元素，并用它们来创造新的雕塑实践。
我拥有多年使用回收和升级再造材料工作经验，从广泛的文化、历史和思想路径中汲取灵感，在自然美学与工业化人类的功能世界之间架起了一座隐秘的桥梁。

Autumnal Melodies
100% silk, wool, polyester fiber, sisal grass fiber, organza, georgette

秋日旋律
100% 真丝，羊毛，聚酯纤维，剑麻草纤维，欧根纱、乔其纱

Shi Lili 石历丽

Works with canvas, waste heat and other industrial materials as medium, through thermoplastic, drop glue solidification, color, color fabric reengineering, in the form of flowers, in order to arouse the sustainable life, sustainable use, clothing sustainable thinking to express all the cycle is the concept of natural rhythm.

作品以帆布、废弃热缩片等工业材料为介质，通过热塑性、滴胶凝固、积色、上色等面料再造手法，以花的形态再现衣的形态，以期唤起对生命的可持续、物用的可持续、服饰的可持续的思考，以表达所有的轮回都是自然的律动之理念。

The Sustainability of Flowers
Canvas, heat shrink film, drip gum, etc

花之可持续
帆布、热缩片、滴胶等

Shi Mei　　　　　　　　　　　　　石梅

Feathers: Feather, wings, the symbol of freedom, courage and fly
Song: music tones, music composition
The work is formed with the wraps of the used coir raincoats gathered from the Jiangnan region("Jiangnan region" is a geographic area in China referring to the region south of the lower reaches of the Yangtze River). The work was sewed up with different types of thread. The threads are ranging as if the streams of life energy disperse and converge. They flow at their own wills till the metamorphosis comes. It's the time to stretch the wings and soar, to achieve its sacred, holy ideal. Just like the paragraph of Schiller in Symphony No.9 *The Ode to Joy* written by Beethoven. "No man must stand alone with outstretched hands before him."
We don't know whether it's the music that makes the coir raincoat stretch out its wings, or the spread of the wings that plays the movement of melody.

羽：羽毛、羽翼，翅膀，象征自由勇敢与飞翔
曲：音调、乐曲
作品利用从江南民间搜寻到的旧蓑衣的披肩部分（江南是中国的一个区域，是指长江下游部分）。本作品用多种材质的线丝穿行缝制，游走的线条有如一股股生命的能量分散汇集，随意流淌，直到完成它的蜕变，撑起了翅膀去飞翔，追随它圣洁的理想，正如贝多芬第九交响曲《欢乐颂》中席勒的篇章："在上帝的羽翼下全世界的人们团结得像兄弟一样。"不知是音乐使蓑衣展开了翅膀，还是羽翼的膨起奏响了旋律的乐章。

A Song of the Feathers
Palm, various copper wire and twine, gauze, silk dress

羽中曲
棕、多种铜铁丝与麻线、纱网、丝衣

The act of humans dressing and undressing is the foundation of this body of work. There is something organically ritualistic to the process of adding and removing clothing to and from the body. In an era of post-humanism, the avatar is a representation of the human within the digital world. Without telling AI how to dress and undress by itself, how could they dress or undress? I began exploring this notion in Clo3d, Maya, and 3D scanners, creating intentional glitches to see how AI would respond and visualize their dressing behavior. The design is inspired by the Apocalypse of the Post-humanism Era. The works present a futurism that is decayed, broken and bordering on zero. Both human and digital forms are explored within a virtual world.

Every piece is patterned and simulated in CLO3D, whilst fabrics/materials are harmonious with the human touch where craft and traditional processes have been explored to create futuristic proposals for physical materials. The base fabric is wearable such as wool and cotton. With post-produced textile design include digital printing, metal coating, laser burning and heating into composited new material.

人类穿衣和脱衣的行为是这一系列作品的基础。在身体上添加和脱掉衣服的过程有一些有机的仪式感。在后人文主义时代，赛博格化身是数字世界中人类的代表。不告诉人机自己如何穿衣服和脱衣服，他们怎么能穿衣服或脱衣服呢？我开始在 Clo3d、Maya 和 3D 扫描仪等设备中探索这个概念，故意制造故障来观察 AI 的反应并可视化它们的着装行为。该设计的灵感来自后人文主义时代的启示录。这些作品呈现出一种腐朽、破碎、接近于零的未来主义。

每件作品均在 CLO3D 中进行图案化和模拟，而织物 / 材料则与人性化和谐，探索工艺和传统工艺，为物理材料创造未来主义方案。底布都是具有可穿戴舒适性的羊毛、棉等基本面料。后期面料处理工艺包括数字印花、水浸电镀金属涂层、激光灼烧和加热成复合新材料。

Apocalypse of Post-humanism
Elastic digital printing mesh, wool, metal coating silk, metal yarn nylon, mesh, cotton, satin, metal coating leather

后人类主义启示录
弹力数码印花内衣网纱、羊毛呢、金属涂层纱、金属丝尼龙、网纱、棉、色丁、金属皮

Sun Xiaoyu 孙晓宇

"*Wu Ji*" series with gray, white gradient tone interpretation to express the theme of the work. "Wuji" is infinite, that is, infinite possibilities, which is a more primitive and ultimate state than Tai Chi. Chinese civilization has a long history and can not be exhausted. Expressing its profound connotation through clothing design works and analyzing Oriental aesthetic thoughts three-dimensional modeling are the inspiration and driving force for the creation of this series of clothes.

The material with natural elements such as cotton and hemp is transformed and enriched by the method of fabric reconstruction, forming new textures, folds and relief effects, and matching with color-free fabrics, producing a distant, ethereal and elegant artistic conception. In the clothing production process, the use of solid clothing skills, the ingenious use of sewing stitches, strengthen the performance of the structure in the clothing, but also make the clothing form in the space display more vivid, full and clear.

《无极》系列以灰、白渐变色调诠释表达作品主题。"无极"便是无穷、即无限可能性，是比太极更加原始、更加终极的状态。中华文明源远流长、不可穷尽。用服装设计作品表现其深远内涵、立体造型的手法解析东方美学思想，是这系列服装创作的灵感与动力。

作品用面料再造方法把具有棉、麻等天然元素的材料改造并丰富，形成新的纹理、褶皱与浮雕效果，并与无色系面料相搭，产生出幽远、空灵、雅致的清新意境。在服装制作工艺中，运用扎实的服装功底，巧妙采用缝纫线迹的效果，强化了结构在服装中的表现作用，也使服装形态在空间展示上更加鲜明、充分、清晰。

Wu Ji
Cotton, hemp

无极
丝棉、丝麻

The *Folding Cities* series is inspired by Chongqing during the epidemic period, where people are isolated at home and their activities are reduced, making the whole city seem to be folded. The intricate route and lighting are the most distinctive urban language of Chongqing. The main design elements of the route map of Chongqing city make clothing a medium reflecting the background of The Times. Clothing accessories for the production of touch color changing light group device, with interactive, through touch sensing or remote control light and change color. The lights come on and the map lines are more clearly visible, symbolizing the city's recovery.

The clothing design research focuses on the regional cultural roots in combination with modern science and technology, fashion design of man-machine interactive, interdisciplinary new exploration and innovation, to create a different paradigm of wearable art design possibilities.

《折叠城市》系列灵感来源于疫情时期的重庆，人们居家隔离，活动范围缩小，整个城市仿佛被折叠。错综复杂的路线及灯光是重庆最有特色城市语言，作品以重庆城区线路图为主要设计元素，使服装成为反映时代背景的媒介。服装外部配件为制作的触控变色灯组装置，具有交互性，可通过触摸感应或遥控发光及变换颜色。灯光亮起，地图线路更加清晰可见，象征着城市的复苏。

本次服装设计作品研究的重点是借地域性的文化根源结合现代科技，实现时尚设计的人机交互性、跨学科性和创新性的新探索，为创造出不同范式的可穿戴艺术设计提供可能性。

Folding Cities
Mixed Media

折叠城市
综合材料

Tu Miaomiao 涂淼淼

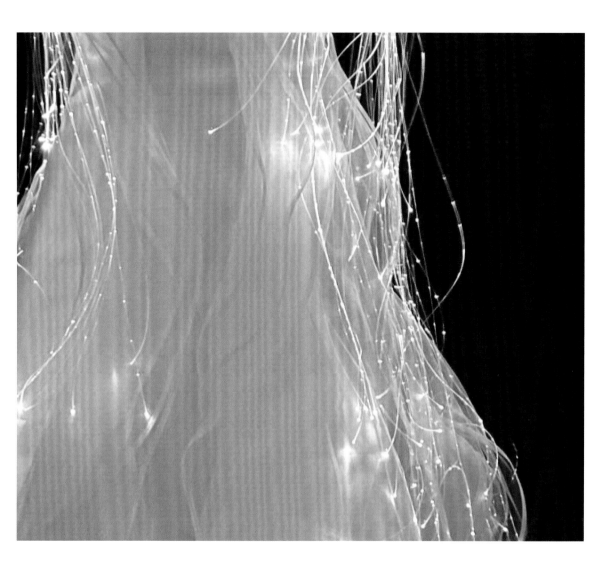

The theme of this design is *"Remolding"*, which means breaking the inherent boundaries, looking for breakthroughs and pursuing innovation. This clothing design is based on this, on the basis of the original dress, the use of new techniques, new materials to design innovation, this work of the traditional dress romantic elegance, but also into the contemporary elements of science and technology, with a sense of the future of science and technology. The fabric is mainly combined with satin and mesh, with fishbone and optical fiber. In the dark environment, the optical fiber silk hangs down the skirt, which has visual impact and fashion without aesthetic feeling.

本次设计的主题是《重塑》，意味着打破固有的界限，找寻突破，追求创新。本次的服装设计基于此，在原有的礼服基础上，运用新的手法，新的材料进行设计创新，既有传统礼服的浪漫典雅，也融入了当代科技元素，具有科技未来感。面料主要采用缎面和网纱进行结合，加以鱼骨、光纤，在暗部环境下光纤丝顺裙褶垂下，极具视觉冲击，时尚又不失美感。

Remolding
Silk, organza, fish bone, mesh yarn, optical fiber

重塑
真丝、欧根纱、鱼骨、网纱、光纤

Wang Lei/Li Qiu 王雷 / 李秋

This work uses sticky dust paper to collect dust from friends' clothes as creative material. Extracting the individual attributes of warmth, scent, and the natural bestowment of dust, it is then folded into angular scales and synthesized into a common piece of clothing. The "Patchwork Garment," composed of dust and new scales from hundreds of people, carries the meanings of blessing and protection. Through the collection and rearrangement of dust, the work narrates touching stories related to "Puyuan Silk".

作品以粘毛尘纸粘取身边朋友衣服上的毛尘为创作材料，提取带有个人属性的温度、气味以及自然给予的尘埃，再折叠成带尖角的鳞片，合成一件生活中常见的衣服。数百人的毛尘与新鳞片组合而成的"百衲衣"具有祈福、保护之意。作品通过取尘的收集与重组，讲述生活中那些与"濮绸"有关的动人故事。

Collecting Dust
Sticky dust paper, lint from clothes

取尘
粘毛尘纸、衣服毛尘

The work is named *"Delight"*, which expresses the emotion and vitality of keeping the heart quiet and joyful under the complicated and chaotic environment. Through 3D digital design, 3D printing and other technologies combined with clothing design, three-dimensional cutting, exploring more possibilities of fashion design methods.

作品取名《心悦》，表达在复杂纷乱的环境下，保持内心安静喜悦的情感与生命力。通过三维数字设计、3D 打印等技术与服装设计、立体裁剪相结合，探索服装设计方法更多可能性。

Delight
3D printing materials, silk, polyester fibers

心悦
3D 打印材料、丝、聚酯纤维

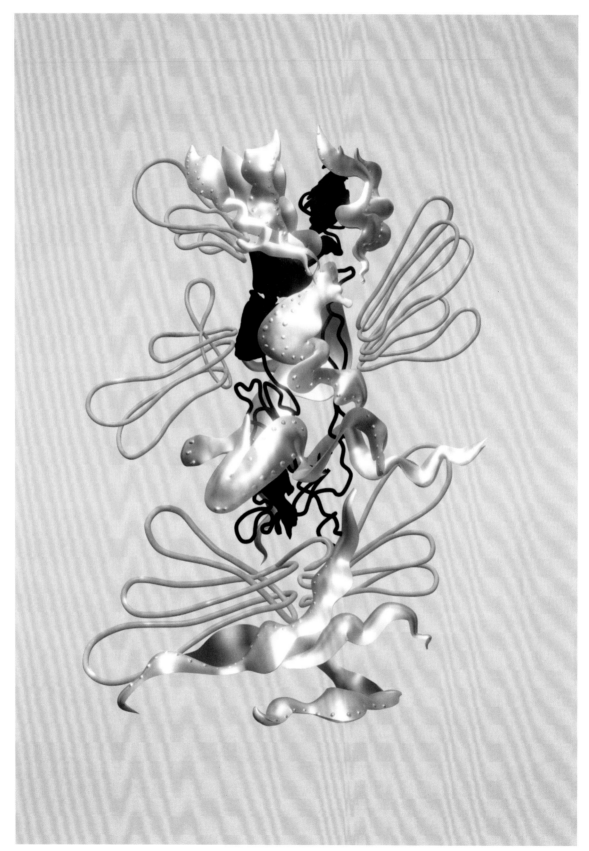

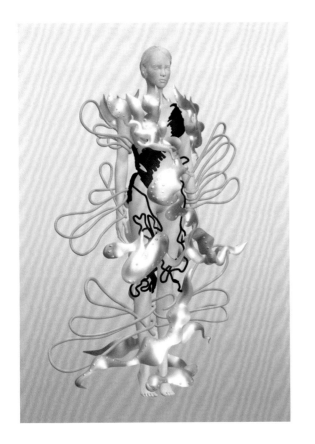

The imagery of the open sea carries the curiosity and vigilance of human beings toward the unknown. This digital fashion work is born in the context that art creation is undergoing great changes brought by the development of digital technology and artificial intelligence, converging and taking shape amongst the dense water plants and hidden drifts.

In Greek mythology, Aphrodite, the goddess of beauty, was born in the sea, and in the blue sea of the Internet, there is a whole new possibility of how beauty can be created and defined. In the work, metal and cables twist and cover the wearer's body like seaweed, and float together with the surrounding halos, as if they are in the deep sea.

Aphrodite of Open Sea comes from the reflection on AI's intervention in art creation, a huge amount of images trained by algorithms to shape a new work of art, for human creators it is a helping hand as well as a challenge, we need to think about how to break free from the unknown dense algae, to realize the benign cycle of what we have taken from ourselves and used for ourselves.

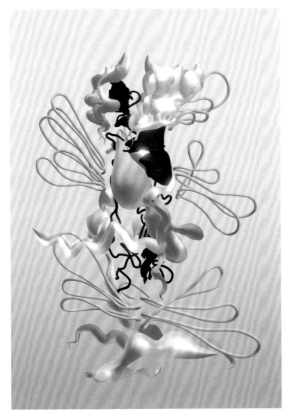

开阔海域的意象，承载着人类对未知领域的好奇和警惕，本件数字时装作品诞生于数字技术和人工智能正在为艺术创作带来巨变的背景下，在茂密的水草和暗流间汇聚成型。

在希腊神话中，美神阿芙罗狄忒诞生于海洋，而在互联网的蓝海里，美如何被创造和定义有了全新的可能。作品中，金属和电缆如海藻般缠绕、覆盖在穿戴者的躯体上，与四周环绕的光圈一同漂浮，仿佛置身于深海之中。《公海里的阿芙罗狄忒》来自对 AI 介入艺术创作的反思，海量图像经过算法训练去塑造全新的艺术作品，对于人类创作者而言是助力也是挑战。我们需要思考如何挣脱未知的密藻，实现取之于己、用之于己的良性循环。

Aphrodite of Open Sea
Virtual works/Video format

公海里的阿芙罗狄忒
虚拟作品 / 视频形式

The design inspiration comes from the Puyuan Town with more than one thousand years history . The design draws on the architectural characteristics of the Jiangnan region during the Song Dynasty, incorporating water corridors and plants to create a unique cultural atmosphere. The pattern design is inspired by famous ancient pagodas, temples, and bridges. Through design, the scenery of Puyuan Town is constructed into a continuous pattern. To realize the concept of sustainability, we adopted environmentally friendly digital printing technology to present the cultural features of Puyuan Town in a more eco-friendly way. In the design, mesh fabric is used and combined with layered design techniques to express the long history of Puyuan Town and the flowing sense of Jiangnan water town. This design work shows the unique beauty of the new Chinese style, while conveying the important values of sustainable development. Combined with environmentally friendly digital printing technology, it not only reduces the dependence on the chemical substances used in traditional printing processes, but also reduces energy and water consumption, aiming to promote the design concept of sustainable development. Therefore, this design is not only a tribute to the unique culture of Puyuan Town, but also hopes to arouse social attention to environmental protection, encourage the public to live in a more sustainable way, and strive to create a better future.

Harmony Bridge Over the Flowing Waters
Cotton, polyester digital print mesh fabric

设计作品灵感源自千年古镇——濮院镇，设计借鉴了宋代江南建筑特色，其中融合水廊和植物，以营造独特的文化氛围。图案的设计灵感来自著名的古塔、寺庙和古桥，通过设计将濮院镇的景色构建成四方连续图案。为了实现可持续的理念，我们采用了环保数码印花技术，以更环保的方式和服装结合以呈现濮院镇的文化特色。在设计中，使用了网纱面料并结合富有层次感设计手法，以表达濮院镇悠久的历史和江南水乡的流动感。这套设计作品展现了独特的新中式风格之美，同时传达了可持续发展的重要价值观，结合环保数码印花技术，不仅减少了对传统印花工艺中使用的化学物质的依赖，还降低了能源和水的消耗，意在推动可持续发展的设计理念。因此，本设计不仅是对濮院镇独特文化的致敬，也希望可以引起社会对环境保护的关注，鼓励大众以更可持续的方式生活，为创造美好未来而努力。

流水廊桥
棉、涤纶数码印花网纱

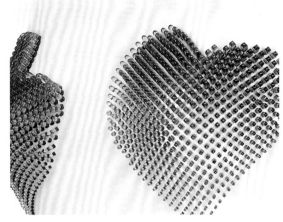

Inspired by the *Tao Te Ching*, "Harmonize with the light, harmonize with the dust", combined with the background of the current era and the economic situation, the release of the trend theme "Harmony of light and dust", intended to reflect the apparel industry in line with the development of the times, in the trend of the times continue to change and innovate, and natural co-prosperity of the fashion concept.

作品灵感来源于《道德经》中的"和其光、同其尘",结合当下时代背景和经济形势,发布趋势主题"和光同尘",意在体现服装行业顺应时代发展,在时代潮流中持续变革创新,与自然共荣的时尚理念。

Harmony of Light and Dust
Fur, 3D printing-resin, polyester, spandex, synthetic fiber

和光同尘
裘皮、3D打印—树脂、涤纶、氨纶、人造纤维

Ji, which means continue, accomplish. It means that only through persistence and continuity can we achieve something. It symbolizes the cycle of people, culture and skills, calling on modern people to return to the natural life and the true nature, and conveying the harmonious coexistence of man and nature, man and society, and man and things.
The work "Ji·Continuation" uses summer cloth as the creation carrier, inspired by the summer cloth performance hemp technology. In the creation, the traditional handmade summer cloth material and dyeing process is improved, and the inter-color performance of hemp technology is innovated, which produces unique color and texture changes in the summer cloth through color weaving.
In the design of dresses, the elements of the appliqued skirt from the horse-faced skirt of the Ming and Qing dynasties are combined with the Western three-dimensional tailoring structure, and the minimalist silhouette and lines are used to express the elegant state of calmness, rationality and naturalness of traditional Chinese aesthetics; in the design of ready-to-wear garments, we emphasize on the principles of saving and clever use, and combined with the virtual clothing technology of CLO3D, we combine the traditional cross-cutting method of China with the modern fashion modeling, and design daily clothing for the traditional narrow summer fabrics. The design innovation of daily apparel for traditional narrow summer cloth fabrics highlights the functional characteristics of summer cloth fabrics, advocates the saving of natural and social resources, tries to translate the Chinese tradition into contemporary language, and interprets the traditional beauty in a modern way of wearing.

Ji·Continuation
Virtual works/Video format

绩——续也，成也。意为坚持而续，方才能有所成绩。寓意着人、文化、技艺的循环更迭、周而复始，呼唤现代人对自然生活和本真的回归，传递人与自然、人与社会、人与物的和谐共生之道。
作品《绩·续》以夏布为创作载体，灵感来自夏布的绩麻工艺。创作中改良传统手工夏布原料及染色工艺，创新间色绩麻工艺，通过色织使夏布产生独特的色彩和肌理变化。
礼服设计运用中国明清时期马面裙中的阑干裙元素与西式立体剪裁结构相结合，用极简的廓型和线条，表达了中国传统美学中的沉静、理智、自然的优雅状态；成衣设计上讲求节用和巧用原则，结合CLO3D虚拟服装技术，将中国传统十字剪裁方式与现代服装造型相结合，针对传统窄幅夏布面料进行日用服饰的设计创新，突出夏布面料自身的功能特性，提倡对自然资源和社会资源的节约，尝试把中国传统转译成当代语言，并以现代的穿着方式演绎传统美。

绩·续
虚拟作品/视频形式

Flowers are one of the important members of nature, or brilliant colors, intoxicating flowers, or plain and indifferent appearance, in our memory and thinking imaging. In the limited thinking bearing, the author uses the artistic expression form of silhouette and the technical method of cloud embroidery to express it in the fashion design, so as to interpret the understanding and expression of flowers. There are various ways of expression, one of which is visual art expression. After the cognition of the object, the image is reconstructed through thinking, and the output is a general graphic art, which is a medium of communication art and an important carrier of information.

花卉是自然界中的重要成员之一，或是绚丽的色彩、醉人的花香，抑或是朴素淡然的外表，在我们的记忆思维中成像。在有限的思维承载中，作者用剪影的艺术表现形式和拨云绣的技艺方法将其表现出来，呈现在服装设计中，以此来诠释对花卉的理解与表达。表达的方式具有多样性，视觉艺术表达是其一，对物象认知后，通过思考而重构成像，并以概括的图形艺术输出，是交流艺术的媒介，亦是信息的重要承载体。

Flowers in Full Bloom
Mulberry silk

繁花盛开
桑蚕丝

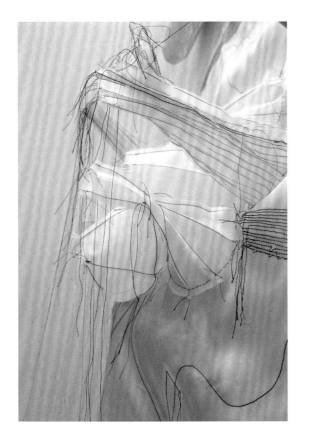

The artwork was inspired by the vases. The shape of a vase is very similar to the curves of a female body, and they are all clever curves. The artist uses the similar characteristics to create works, presenting the 3D and 2D shapes of the objects on the Oriental plane structure clothing, using extended and imperfect stitches to express the sense of ease and carelessness, and making the relationship between the shape of the plane and the object of space; making the plane line echo the space line. The transparent effect of the material, the superimposed effect of multi-layer silk fabric highlights the characteristics of the work to express the beauty of uncertainty, just like the shape of nothing, the line of plausible. The work interprets the artist's understanding of emptiness and agility in Oriental aesthetics.It can be hung and worn, presenting a completely different visual effect.

作品灵感源于日常生活中插花的器皿。这些器物的造型与女性身体线条非常相似，皆为灵动的曲线。作者借用两者形似的特性进行创作，将器物的 3D 与 2D 造型呈现在东方平面结构的服装上，并用加长的线迹和不完美的针脚，表达轻松和不经意之感。使平面之器形与空间之器物发生关系；使平面之线与空间之线产生呼应。材质的透明效果、多层真丝面料产生的叠加效果强化了作品所要表达的不确定性之美——若有若无之形、似是而非之线。作品诠释作者对东方审美中的空无与灵动的理解。作品可悬挂、可穿戴，呈现出完全不同的视觉效果。

The Vases
Natural Silk Satin, thread (Black)

花器
原色真丝缎、长纤维缝纫线（黑色）

Haze Ng (HongKong, China)/Kinor Jiang (HongKong, China) 吴国禧（中国香港）/姜绶祥（中国香港）

The creation combines the Hong Kong Cheongsam Making Technique, which is a representative intangible cultural heritage of China, with advanced metallic plating technology, attempting to explore the design possibility of the traditional craft with unique and novel textile materials. The work inherits the classic dress form of Hong Kong men's cheongsam, featuring the seamless shoulders and grown-on sleeves, right opening, round neckline and straight hems of the traditional Chinese robe. The form presents a grand oriental poise as well as a classical refinement. The metallic plating cast an iridescent spectrum on the authentic robe, shimmering with numerous colors under different lightings and angles.

此作品结合中国国家级非遗代表作名录项目"香港中式长衫制作技艺"及高端金属镀覆技术，尝试以创新独特的纺织物料为传统非遗技艺注入创新设计的可能性。作品承经典的香港男装长衫规格，具传统中式袍服的平面结构，连肩平袖、右衽开襟，圆领直裾。其形制恢宏大气，同时古典儒雅。面料的金属镀覆在长袍上谱出虹光，在不同的灯光和角度下金光闪烁，色彩千变万化。

Iridescent Classic
Metal-plated polyester slub tabby

虹霓经典
金属镀覆涤纶竹纹绸

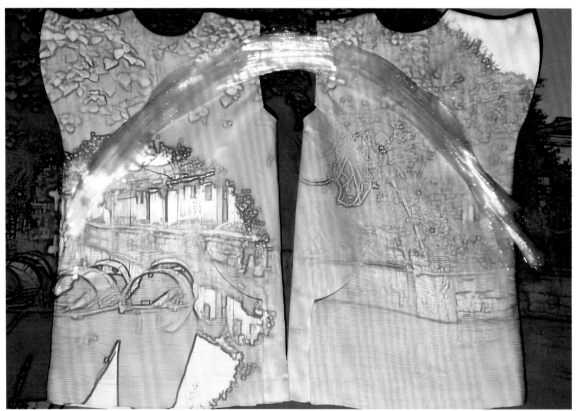

Wu Jing 吴晶

 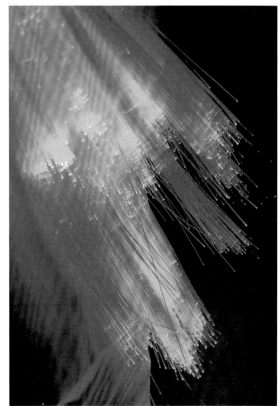

The work is inspired by the small bridges and painting walls of the Jiangnan water town. The slightly upturned shoulders symbolize the beauty of the flying eaves of Jiangnan houses. The designer uses pleated fabrics and fiber optic materials as well as curved structures to express the impression of the ancient stone bridge with a half-bent moon and rippling water in the Jiangnan water town.

作品灵感来自江南水乡的小桥流水粉墙。微微上翘的肩部象征着江南民居的飞檐之美。设计师运用褶皱面料和光纤材料以及弧线结构，来表现古意石桥月半弯与水波荡漾的江南水乡印象。

Jiangnan Impression
Linen, polyester, optical fiber

江南印象
棉麻、涤纶、光纤

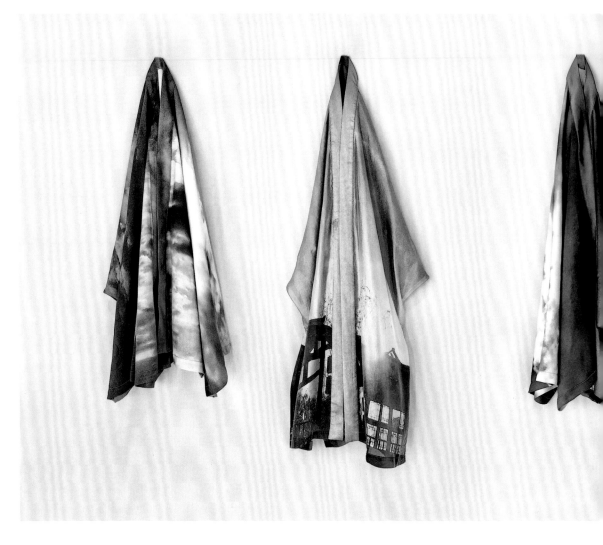
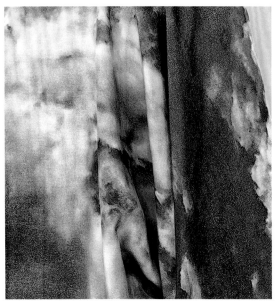

Xiong Yi 熊艺

The "Zen Hall" is a place where you can experience life with every breath and breath, amidst the atmosphere of water and nature. "Zen clothing" is suitable for comfortable and meditative meditation.

Having a heart of belonging is Zen. The sky in 2020, for me, is Strength, Reflection, and Hope. In this series of creations, the author retains the traditional Eastern one-piece cutting concept, tailoring "Zen clothes" and incorporating digital printing techniques to present the sky of 2020. This series is my (author) reflection on the symbiosis of traditional and contemporary ideas. The sky of 2020 interprets the vastness of the universe, the beauty of the sky, and everything in the world is rooted in the heart. Destiny is created by oneself, form is generated by the heart, and all things in the world are transformed into forms. The heart remains motionless, all things remain motionless, the heart remains unchanged, and all things remain unchanged. Having a heart of belonging is Zen.

"禅堂",在水意与自然的气息中,一呼一吸间感受生活。
"禅衣",以舒适随体、禅修打坐为宜。
心有所属,即是禅。2020年的天空,于我(作者),是力量,是反思,是希望。
本系列创作,作者保留东方传统一片式裁剪的创作理念,裁制"禅衣",融入数码印花工艺呈现2020年的天空。本系列是我(作者)一直以来,对传统与当代共生思想的再一次思考。2020年的天空,诠释宇宙浩瀚,天空之美,世间万事万物,皆系于心。命由己造,相由心生,世间万物皆是化相,心不动,万物皆不动,心不变,万物皆不变。心有所属,即是禅。

the Sky in 2020
Silk satin, tiansi linen

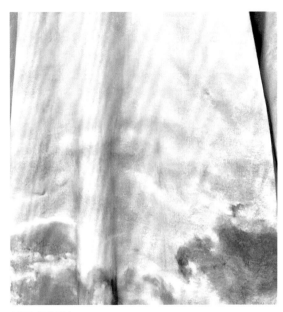

2020年的天空
真丝缎、天丝亚麻

HSU CHIU | (Taiwan, China)　　　　　　　　徐秋宜（中国台湾）

Inspired by the enchanting oceans and vibrant flowers, these elements serve as the wellspring of creativity. I intertwine flowers, the sea, flowing streams, and fantastical elements in my work, creating a dreamy texture. Such an artistic style conjures thoughts of a Mobius-esque fantasy world, where details play a charming role, and this passion leads viewers to the depths of life's beauty. The thematic works are not merely expressions of admiration for natural beauty, they also encapsulate a profound ardor for art.

In terms of creation, the work is inspired by colorful drips and drops, and uses these elements to present a surrealist style, boldly using turquoise, pink and yellow colors. Through the fusion of colors and highlighting of textures, the high contrast of colors enhances the visual effect of the work, creating a dramatic effect of vivid contrasts that resonates with the viewer. This design style is like an interdimensional dreamland wearing fantasy patterned clothing, giving the works a sense of fantasy and science fiction, while presenting the confidence and unique beauty of the exotic world, encouraging people to pursue the infinite possibilities of experimental creation. These charming inspirations and different elements cleverly meet, suggesting a unique world that combines art, virtual, fantasy and reality, allowing people to feel the experimental and fantastical vitality of life from time to time!

Let's embark on a magical journey of fashion, digital innovation and art, and fearlessly delving into the inexhaustible allure they hold together.

在美丽海洋与缤纷花朵的启发下，这些元素成为创作的灵感来源。我将花朵、海洋和滴流，以及奇幻的元素融合于作品之中，营造出梦幻的质感。这样的艺术风格引人联想到莫比乌斯的奇幻世界，细节在其中扮演着具有魅力的角色，这份热情将观者引领至生命美的深处。主题作品不仅是对自然美的赞美，更凝聚了对艺术的深厚热情。

创作方面，作品从多彩的点滴洒落中获得灵感，运用这些元素呈现超现实主义风格，大胆运用绿松石、粉红和黄色等色彩。通过色彩融合和质感的凸显，高对比的色彩增强了作品的视觉效果，塑造了极富生动对比的戏剧性效果，进而使作品引起观者共鸣。这种设计风格仿佛是穿着奇幻图案服饰的异次元梦境，赋予了作品奇幻与科幻的虚幻感，同时呈现了异域世界的自信和独特之美，鼓励人们追寻实验创作无限的可能性。这些魅力灵感与不同元素巧妙地遇合，暗示了一个结合艺术、虚拟、幻想和现实的独特世界，让人感受生活中时而充满实验的、奇幻的生命力！

就让我们携手踏上时尚数位与艺术的奇幻之旅，一同勇敢地探索其中无尽的魅力。

Fantasy Odyssey: "Fashion of the Floating World"
Virtual works/Video format

奇幻奥德赛："浮生时尚"
虚拟作品 / 视频形式

This collection is based on the traditional southern Chinese clothing silhouette, with Western-style fabric craftsmanship used for teaching. It draws visual inspiration from the trends of artificial intelligence, virtual fashion, and the metaverse. This combination of Chinese and Western dress styles allows contemporary design to interpret traditional discourse, possessing more of the romance and sentimentality unique to Chinese style.

With exaggerated shapes and strong visual accessories, it reshapes the independent image of modern Jiangnan women, showcasing a definition of blooming beauty that both inherits tradition and breaks new ground. The attitude of young designers serves as a fashionable carrier for inheriting and continuing traditional culture, with "continuing the ancient as the trend" as the original discourse of the design.

本系列以中国传统的江南服装制式为廓型，以西式手法的面料工艺进行教学。从流行趋势中的人工智能、虚拟时装以及元宇宙获得视觉效果灵感。在中西结合的穿搭方式下，当代的设计能够演绎传统的话语，拥有更多属于中式的浪漫与情怀。

作品以夸张的造型、强烈视觉的配饰，重塑当代江南女子的独立形象，展示继往开来又芳华绝代的定义。用青年设计师的态度去成为继承和延续传统文化的时尚载体，以"续古为潮"作为设计的原话语。

As If the Old Friend Comes
Satin, organza, tencel, decorative pearls, custom-made hardware for bras

似是故人来
缎面、欧根纱、天丝、装饰珍珠、胸衣定制五金

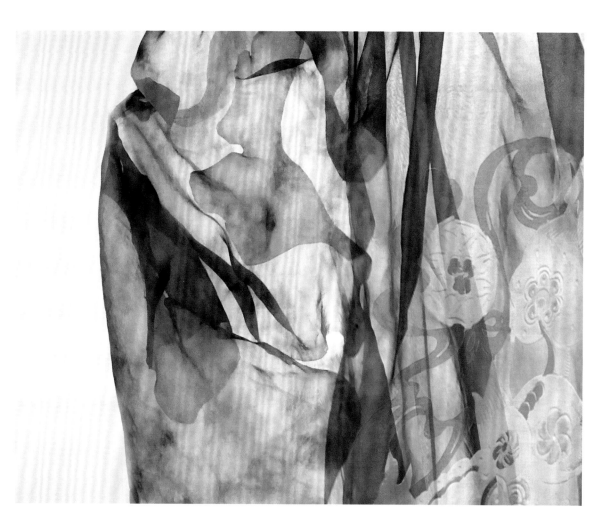

The work is based on the theme of people celebrating the good harvest during the autumn harvest, and uses golden brown, brown, brown-green, deep orange, and yellow-gray series to form an autumn harvest color; folk paper-cut lion dance symbolizes celebrations; vertical and horizontal folds and geometric patterns Collages represent the deep earth that nurtures us.

The long dress of Silk organza, Vegetable Dyeing Technology with black bean clothing and Pu'er tea as the main raw materials, combined with laser cutting and hollowing out technology, it reflects the modern application of traditional textile technology and folk ecological dyeing technology. The work is concise and vivid, smooth and natural, and contains ecological concepts and Chinese elements. It is a design attempt combining modernity and tradition, ecology and folk customs.

作品以秋收时节人们庆祝美好收获为主题，以金棕色、咖色、褐绿色、深橘黄色、黄灰色系列，构成一幅秋季丰收的色彩；民间剪纸狮子舞象征着庆祝活动；纵横的褶皱和几何图案拼贴代表着养育我们的深厚土地。

真丝欧根纱长裙，以黑豆衣、普洱茶为主要原料的草木染色技术，配合激光切割与镂空技术等工艺，体现传统纺织技术和民间生态染色工艺的现代应用。作品简洁生动，流畅自然，蕴含生态理念和中华元素，是一次现代与传统，生态与民俗结合的设计尝试。

Celebrate a Bumper Harvest
Silk organza, laser cutting, collage, pearl embroidery, vegetation dyeing

庆丰收
真丝欧根纱、激光切割、拼贴、刺绣、草木染色

The millions of hands represent the multitude of human beings. Hand in hand, the dance of life goes on without end. Together, regardless of the ups and downs, all of us, hold the moonlight in our hands and travel freely in heaven and earth.

千万只手是芸芸众生之隐喻。手手相依的生命之舞从不停息。无论跌宕起伏，众人一道手捧月光，畅游天地。

Water-Moon in Hands
Polyester

掬月游
涤纶

During the initial stages of creation, I conducted experiments to produce a variety of fully biodegradable plant-based fabrics, including options like "Thousand Flowers Fabric," "Lemon Peel Fabric," and "Grapefruit Peel Fabric." However, for the final garment, I utilized discarded coffee grounds fabric. This choice was motivated by the fact that coffee grounds, as a raw material, represent a common form of biomass waste. Hence, I aimed to transform them into a biogenic fabric, attempting to reduce waste production.

The work's creative inspiration is the ceaseless dance of plants. The body is in flux, becoming organic, a structure without fixed organization, aiming to rouse people's bodily consciousness through the perception of plants, and to unfurl constricted bodily perceptions.

我通过实验，在创作初期制作了很多可全降解的植物面料，例如，千花面料、柠檬皮面料、柚子皮面料等。但是，在最后的成衣作品中使用的是废弃的咖啡渣面料，因为原材料咖啡渣是一种常见的生物质废弃物，所以我想尝试将其转化为生物面料，以此尝试减少废弃物的产生。

作品的创作灵感是植物的不息之舞。身体是流动的，去有机化的，一种无固定组织的身体，试图通过植物的感知来唤醒人们的身体意识，舒展闭塞的身体知觉。

Dance Dance Dance
Fully degradable coffee grounds fabric, bamboo strips (support structure)

舞 舞 舞
可全降解咖啡渣面料、竹条（支撑结构）

Intersection and weaving are one of the main characteristics of harmonious coexistence between humans and nature, and interweaving and crossing boundaries are also important concepts in modern design. This design focuses on the theme of "communication and weaving", highlighting the interweaving of "etiquette" and "fashion" in design concepts. The interweaving of three-dimensional decoration and multiple material reconstructions in terms of expression methods. The materials are interwoven with plain woven fabric and mesh woven material, combining three-dimensional design elements with the elegance and luxury of Western style long dresses, showcasing modern women's dignified atmosphere, vitality and fashion, as well as positive and upward personality style.

Communication and Weaving
Chemical fiber fabrics, mesh, transparent beads, metal wires, rice beads, etc

交与织是人与人、人与自然和谐共存关系的主要关系特征之一，交织与跨界也是现代设计的重要理念。本设计以"交·织"为主题，突出设计观念上"礼"与"尚"的交织；表现方法为立体装饰与多种材料再造的交织；材料运用为平纹梭织面料与网纱编织材料的交织等，将立体设计元素与西式长礼服的典雅华贵相结合，展现出现代女性端庄大气、活力时尚以及积极向上的个性风采。

交·织
化纤面料、网纱、透明珠子、金属丝、米珠等

Tim Yip (HongKong, China)　　　　　　　　　　　　叶锦添（中国香港）

This conceptual outfit is part of Tim Yip's *Cloud* series which addresses important topics in society, using waste materials to evoke reflection on consumerism, environmental changes and the future. *Phone* uses 23 mobile phones and luminous strips to convey the anxiety of receiving constant information and the pressures of understanding one's identity within an increasingly interactive society. We hear different messages from everywhere on our phones. But in the ocean, they become messengers without messages.

About *Cloud*

"The ocean always has deep memories. I saw a lot of different items of rubbish piled up on the beach, on the edge of the sea. These things only have their shape without functions and basic names there. I found a lot of clues during that period I spent by the ocean. 25 different stories slowly emerged from the contemporary collective consciousness. I invited 20 creative designers in London to collaborate with me to make each story into a costume. At the same time, I interviewed 50 London teenagers to find out who they are, and what they think the world will look like in 1000 years. Through their multiple views, I found that in one city there are living thousands of different souls.

Cloud is like a magic box containing every aspect of human history, and reflecting the future, where our new generation are heading. Those two angles become our content like a mirror. All *Cloud* costumes reflect time.

Our stories come from materials we face daily, reformed as an art installation on the body. So these are not just costumes, but something else unknown. We can use all kinds of things... but all come from the rubbish of mankind. They sometimes come with their similarities, sometimes by chance, sometimes they reform as undefined figures.

Although the world is multilayered, it is simplified by humans as power or money, to mark their position. As a mass you are being told what is happening and what will happen. If you are middle class, you are working for the upper class. The upper class only accounts for a very small percentage of the population. But if you are lucky enough to be among them, you decide what happens and fight for what will happen. This is absolutely how a big company is structured. All by numbers, calculations, relationships, passwords, limited information, entertaining the mass. You are told that you are free, but we are all straightjacketed. *Cloud* has a feeling of memories of freedom. Chaos coexists with that freedom."

Phone
PVC, LED strips, phone dummies, electrical wires

这件概念服装是《云》系列作品的其中一件。该系列以艺术服装从社会热点话题中汲取灵感，以废弃材料为设计原料，试图唤起人们对消费主义、环境变化以及未来的反思。《手机》（*Phone*）这件作品用23台手机和特殊的发光条，诠释了当代急速互动的社会中，人们每天都从手机上收到各种信息时的焦虑，以及找寻自我身份认同时的不安。我们整天都从手机上收到各种信息。但在海洋中，它们成为没有信息的信使。

关于《云》

"海洋永远都流传着深刻的记忆，我在海的边缘看到很多不同的垃圾堆积在沙滩上，这些东西就只剩下它的形状，功能与基本的名字已经不复再有，这些都是我们的曾经，现在已经成为不知名的垃圾。在这期间我找到非常多的蛛丝马迹，25个不同的故事，慢慢浮现在当代意识里。同时，如涓涓细流般诉说着它们的故事，呈现在我们的眼前。我邀请了20位伦敦的设计师，和我一同将每一个故事变成一套服装。这过程中，我在伦敦遇到了许多有创意的人。同时，我采访了50位生活在伦敦的青少年，了解他们来自何方，他们是谁，他们对未来的设想，他们认为世界在1000年后会是什么样子。我察觉了他们的梦想和挫折。通过他们的多重视角，我发现在一个城市中，生活着成千上万不同的灵魂。

《云》宛如一个魔法盒，蕴藏着人类过往的方方面面，映照着新一代人的未来。过往与未来，像一面镜子，成为我们展览的焦点，让每一件'云'服装反映一段岁月。

我们的故事来自人们日常使用的材料，在人身重塑为艺术品，所以它们是超越服装的存在，是一种未知的存在。无论我们用何种逻辑、材料、审美、艺术形式，均来自人类的废弃物。它们的出现有时是伴随着相似物的，有时是偶然的，有时是模糊的。

尽管这是个多层的世界，人类却把它简化成权势与金钱，以展示自身的地位。作为芸芸众生中的一分子，你被需要时就被告知正在发生什么、即将发生什么。如果你身处中产阶级，那么你是在为上层阶级服务，而上层阶级只占全部人口的很小一部分。但是，如果你有幸加入了他们，你就能决定发生什么并且争取即将发生什么，而一家大公司的构造绝对是这样的。

我们都利用数字、计算、关系、密码限制了信息的流通，以取悦大众。你被告知你自由，但是我们都深受约束。《云》传达一种追忆自由的感觉，而那自由与混乱共存。"

手机
PVC、LED灯带、手机模型、电线

Yu Yimeng 余一萌

In the sea, there are bones, where strength and flexibility coexist. This artwork consists of two versions: virtual and tangible. Utilizing 3D modeling and 3D printing techniques, it employs algorithmic generation to transform the magnificent landscapes of the natural world into wearable objects. The supple waves, solid framework, fractal patterns, and natural randomness are synthesized through this process to harmoniously unify opposing elements: the organic unity of yin and yang, straight and curved, rigid and flexible, boldness and elegance, order and chaos. These balanced juxtapositions reflect the creative philosophy embraced by this piece.

海中有骨，刚柔并济。该作品包含虚拟与实体两个版本，使用三维建模技术与 3D 打印技术，以算法生成的方式将自然界的壮阔景观化为身体穿戴物。柔软的海浪、坚硬的骨架、分形的秩序、自然的随机，对立事物的有机统一、阴与阳、直与曲、刚与柔、张扬与优雅、有序与无序的平衡是本件作品秉持的创作哲学。

Sea Bones
Full-color resin

海骨
全彩树脂

Parallel Consciousness, a digital fashion film, is based on the intersection of digital technology, artificial intelligence and fashion. It explores the self-consciousness through the awakening of organic life and the possibility of individual shaping through the generation of inorganic clothing. The film goes back to the origin of the universe, the explosion of the black hole causes the creation of time cracks, and the waking individual is endowed with the ability of independent thinking and perception by the second skin, and then prays into the other selves sleeping in the parallel universe. In the virtual world, the evolution of ideology, identity and self-expression derived from fashion as a touchpoint is at the heart of the film's exploration.

《平行意识》数字时尚影片，立足于数字技术、人工智能与时尚领域的交叉点，以有机生命的苏醒展开对自主意识的思辨，以无机服装的生成探讨个体塑造的可能。影片回溯至宇宙的起源，黑洞的爆炸致使时间裂缝的产生，苏醒的个体被第二层皮肤赋予了独立思考与感知的能力，进而窥探到平行宇宙中沉睡的其他自我。在虚拟世界中，以时装为触点衍生的意识形态、身份认同以及自我表达的演变，是本片探索的核心。

Parallel Consciousness
Virtual works/Video format

平行意识
虚拟作品 / 视频形式

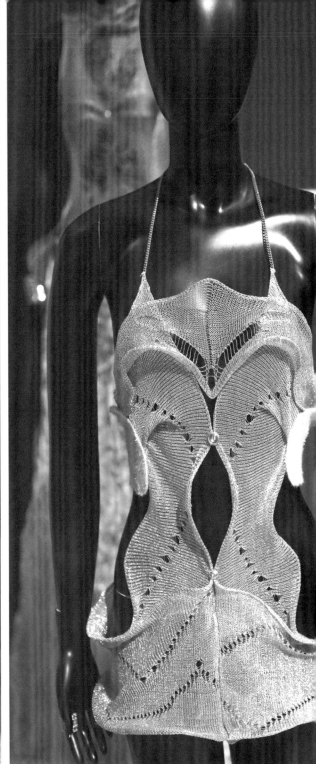

Yu Chenxi 庾晨溪

Knitting art has a high degree of plasticity and exploration possibilities, and knitting the types, ranges and artistic concepts of clothing are also diversifying development. According to the above context, inspired the author for the knitted clothing in practical. So thinking of the value meaning of the impractical field. Works under the tide of the times, starting point of the situation is the loss of the sense of time in the big environment. The body of the life is lived in a short-term cycle and extracted from the art of process the idea. *7 Days* represents the cycle of a week, where everything is made up of one stitch a ray of the composition of this process represents the infinity and confusion of time, and at the same time should right to the process of suffering or the way of accumulating goals, and this repetitive behavior is what I have experienced most in the corresponding era.

针织艺术都有着极高的可塑性和探索可能性，而针织服装的类型、范围及艺术观念也随之朝着多元化的方向发展。根据上述语境，启发了作者对于针织服装在实用及不实用领域的价值意义的思考。作品以时代洪流下的处境为出发点，在大环境下产生对于时间感的迷失，体验到生活在短期性的轮回中度过，并从过程艺术中提取观念。《七日》代表一周的循环，所有东西都由一针一线组成，这个过程代表了时间的无尽与迷乱，同时应对了过程对煎熬又或者是积累目标的方式，而这样重复性的行为就是我在相应时代下体验最深的东西。

7 Days
Mixed media

七日
综合材料

Yuan Yidan 原一丹

The work apply "biological images" and "biological concepts" to artistic creation. Based on the relationship between the modular structural characteristics of insect bodies and the human body, the subject consciousness of insects is explored from the perspective of virtual fashion, and the philosophy of "dynamic balance" in life is considered. If human bodies and insect bodies are emotionally interlinked, should we change the way we treat them? Or give life more free space, so as to start a new definition of the relationship between "non-human" life forms.

作品将"生物图像"与"生物学概念"应用于艺术创作。基于昆虫身体模型化的结构特点与人体产生关系，在虚拟时尚视阈下对昆虫的主体意识进行探究，思考生命体中"动态平衡"的哲学性。如果人类身体与昆虫身体通过依赖所构建的生物体具有情绪互通性，那我们应该改变对待它们的方式吗？或者赋予生命更多自由空间，从而展开对"非人类"生命体关系的新定义。

Insects Fantasy
Virtual works/Video format

昆虫幻想
虚拟作品/视频形式

Zeng Fengfei 曾凤飞

This collection of ZENG FENGFEI is based on the theme of *Encounter*. Taking inspiration from the land of Bamin (Fujian Province), you can feel the natural beauty in this collection, as if taking a trip close to the landscape. In terms of creation approach, this collection draws on the charm of calligraphy and literati painting. Through the fabrics and decorations full of romantic colors, light and shadow, dots, lines, and surfaces, a gorgeous picture of the world is outlined.
The materials used for this collection are fabrics full of natural patterns, textured silk-polyester blended fabrics, complemented by technological materials; totems of oriental elements such as natural flowers, flowing clouds and water are embroidered or jacquard weaved in the garments. Taking the details of the landscape paintings of Bamin, this collection is young and contemporary with the addition of innovative fabrics, injecting multiple aesthetics into the new products under the classic structure.

曾凤飞该系列作品以《际遇》为主题，以八闽大地为题材，可以畅享服装设计中的自然美景，畅快呼吸，开启融入自然、亲近山水的旅行。在创作手法上，系列作品汲取书墨风韵和悠畅寄怀的文心画意，闲居理气，化合为心。透过充满浪漫色彩的面料与装饰，光与影、点线面构图的画面，裁剪出一幅幅大千世界的绚烂画面。
作品材料为大量充满自然图纹的面料，具有质感的丝涤混纺面料，辅以科技感的材质、图腾以及以自然花卉和行云流水之姿，将东方元素时而刺绣时而提花融合于作品中，撷取具象八闽风采山水画细节，以年轻当代的形式穿插于异材质面料，为经典架构之下的系列新品注入多重美感。

Encounter
Mulberry silk, polyester

际遇
桑蚕丝、聚酯纤维

Zhang Gang 张刚

Inspired by the water towns in the South of the Yangtze River, the works are decorated with round pieces of different colors, combined with traditional hand-stitched techniques, to create a hazy view of the misty rain south of the Yangtze River, aiming to express the author's deep memory of the misty rain scene in the South of the Yangtze River.

作品以江南水乡为灵感,用不同色彩的圆片做装饰,结合传统手工针缝工艺,营造出朦胧的烟雨江南景色,旨在表达作者对江南朦胧烟雨情景的深深记忆。

Hazy Jiangnan thick memory
Xiangyun gauze, real silk, organza

朦胧江南浓浓记
香云纱、真丝、欧根纱

The work uses silk gauze as thin as a cicada's wings and silk with delicate texture as materials, injects unique texture and color into each piece of silk gauze through ancient hand-dyed techniques, and uses techniques such as reorganization, stacking, and crossing to establish materials, spaces and the staggered collision between the pattern structure creates a unique and avant-garde series.

作品用薄如蝉翼的真丝绡和质感细腻的蚕丝为材料，通过古老的手工染色技艺，为每片真丝绡注入独特的纹理和色彩，运用了重组、堆叠、交叉等手法，建立材料、空间与图案结构之间的交错碰撞，创造出独特而前卫的系列作品。

Amber Silk Spring and Autumn
Silk gauze, silk

琥珀蚕丝渡春秋
真丝绡、蚕丝

Zhang Peng   张鹏

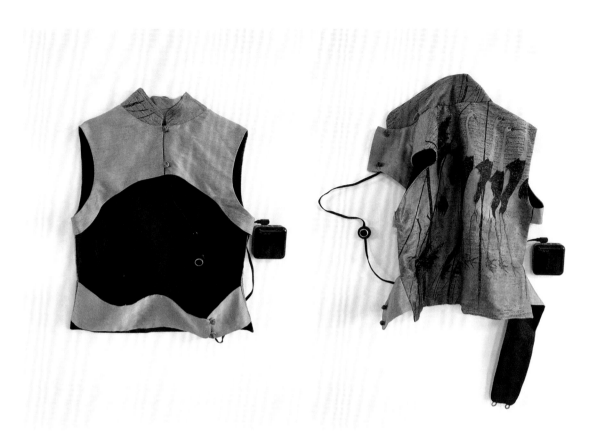

The profound and long history of Chinese culture has left us with countless cultural treasures. Clothing culture is the most intuitive carrier of traditional Chinese culture, and TCM (Traditional Chinese Medicine) culture is the crystallization of the wisdom of the Chinese nation, which contains the philosophical wisdom of the Chinese nation over thousands of years and people's yearning for a healthy life.

This series of clothes takes "Walking TCM" as the theme and traditional clothes as the carrier to examine the medicine clothing method and thermal therapy of TCM, and strives to integrate the cultural characteristics of TCM in the design of materials, structures, and functions, so as to conduct an exploration of cross-border integration in the clothing healthcare design.

中华文化博大精深、源远流长，为我们留下了无数的文化瑰宝。服饰文化是中华传统文化最直观的载体，而中医药文化更是中华民族智慧的结晶，蕴含着中华民族几千年的哲学智慧和人们对于健康生活的向往。

本系列服装以《行走的中医》为主题，以传统服饰为载体，考量中医药衣法与温热疗法，在材料、结构、功能的设计中都力图融入中医药文化特色，进行服装康养设计的跨界融合探索。

Walking TCM
Hand-woven, gambiered canton gauze process tussah silk, dupion silk, spun silk, volcanic fabric, nano-heating flexible material

行走的中医
手工织造、香云纱工艺柞蚕丝、双宫丝、绢丝、火山岩面料、纳米加热柔性材料

Zhang Tingting 张婷婷

The work is completed by the intangible Cultural heritage skill-wool wet felting process, inspired by the Chinese blue and white, using three colors of blue for haloing the charm of ink, exaggerated and enlarged pattern outlines the beauty of the female form of the spirit of movement.

作品由非遗技艺——羊毛湿毡工艺完成,灵感来源于中国青花,采用三色蓝晕染出水墨的气韵,夸张放大的纹样勾勒出女性形体灵动之美。

The Memory of Blue and White Porcelain
Natural wool (handmade wool felt)

青花记忆
天然羊毛(手工羊毛毡)

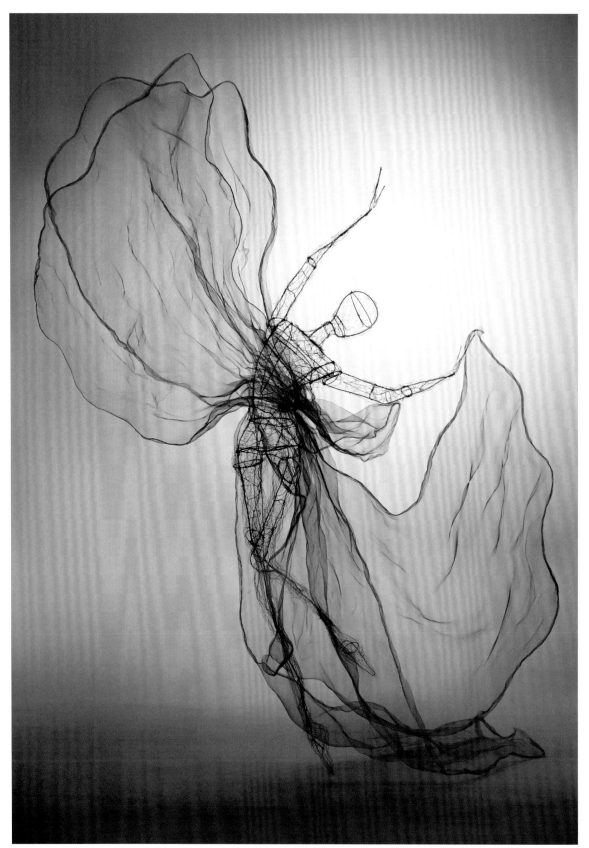

Zhou Zhaohui 周朝晖

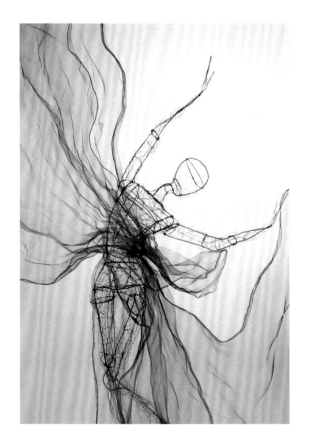 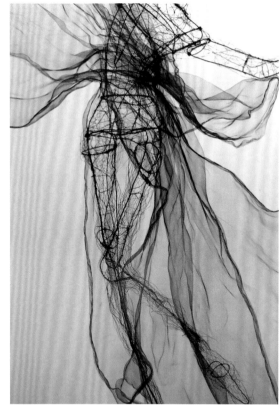

Romantic dresses flying lightly, spirit dancers dancing gracefully. They are not dancing with their body, but with their soul. To awaken people's love for human's clothing, steel wire and steel wire mesh have been used by its designers, elaborately to create a piece of art, which is a harmonious unity of clothing and human beings. How beautiful and dreamy it is!

浪漫长裙迎风飘逸，灵魂舞者轻舞飞扬。他们不是用身体在跳舞，而是用灵魂在跳舞。设计师运用钢丝网和钢丝织造出一件亦衣亦人、亦人亦衣，衣人合一的服装艺术品，形成一种唯美梦幻的感觉。

Soul Dancer
Barbed wire, iron wire

灵魂舞者
钢丝网、钢丝

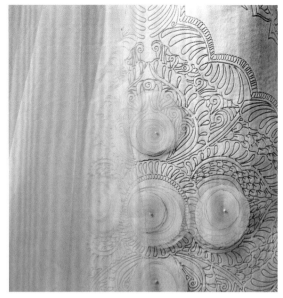

The inspiration for this work comes from the flexibility of water in the water towns of Jiangnan (regions south of the Yangtze River).
It incorporates the flowing sensation of water as well as the water wave patterns from ethnic elements, such as the "Wotuo" pattern.
The artwork uses white as the base color and combines modern laser engraving techniques with traditional "Suoxiu" embroidery technology to decorate thick wool felt and delicate silk fabric with water wave patterns. It is further adorned with a necklace-like decoration of white beads of varying transparency, creating a fashion art piece titled "*Between Flowing Waters*: *A Dialogue between Modernity and Tradition.*"

本作品灵感来源于江南水乡"水"的灵动。
作品将以"水"为介质的流动感和以民族元素中的水纹状"窝妥"纹作为灵感要素。
作品以白色为基础色调,以现代的激光雕刻技术和传统的"锁绣"技艺,分别在厚重的羊毛毡和轻薄的绡上进行水纹的装饰,再点缀以各种透明度的白色调珠子连缀而成的水纹颈饰,完成《盈盈一水间——现代与传统的对视》的时装艺术作品。

Between Flowing Waters
Wool felt, embroidery, beads, fishing line

盈盈一水间
羊毛毡、绡、珠子、鱼线

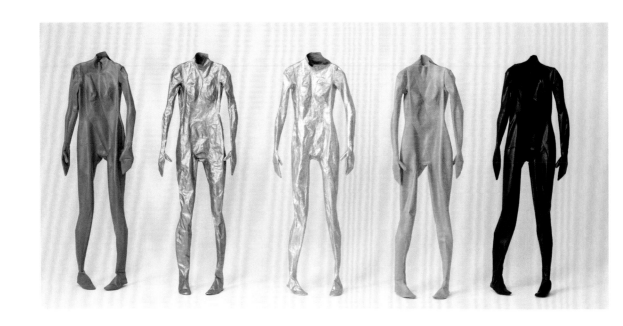

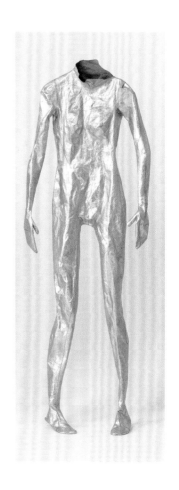

Zou You　　　　　　　　　　　邹游

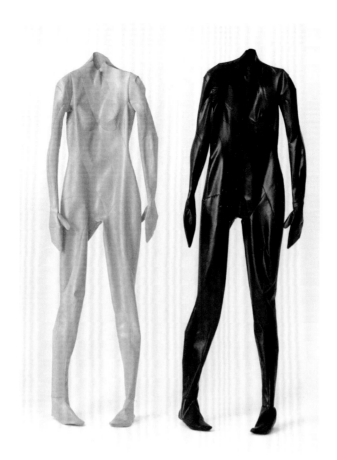

Imagine the image of the person you know? Is it possible to say that the social image of a person is undoubtedly a form of dress? Clothing is like the second skin of the human being, which is completely integrated and inseparable from the body in the identity of the human being. The possibilities of the style of clothing wrapped around the body are infinite, judging from the experience of history. As a creator, there are various possibilities for combining a new material with a new language of wrapping the body, and the result is very personal, so this is one of the exercises.

想象一下，你所知觉的人，是怎样一种形象？
是否可以这样说，人的社会形象无疑都是一种着装的形态？服装犹如人的第二层肌肤，在人的身份设定中，已经和身体完全融凝在一起，无法分割。
服装包裹身体的风格样式，从历史的经验来看，其可能性是无限的。作为创作者，在一种新材料和一种新的包裹身体的语言的组合是有各种可能的，结果也是非常个人的，此为练习之一。

Physical Exercises
Waterproof oxford cloth, dupont paper

身体练习
防水牛津布、杜邦纸

Lyu Yue (Aluna)/Jin Xiaoyao/Tongxiang Yongxin Clothing Co., Ltd.

吕越 / 金小尧 / 桐乡市永欣服饰有限公司

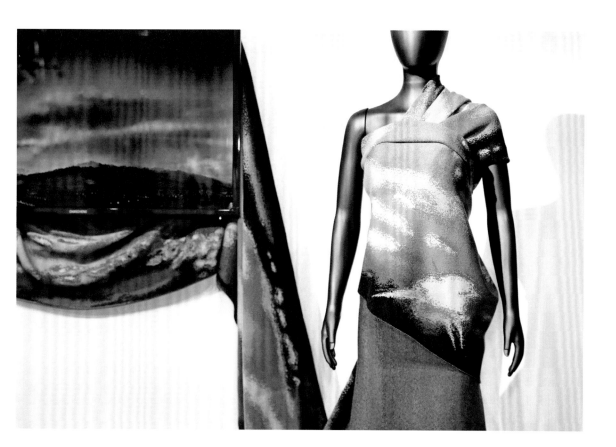

The artist of Sunset Moment, Lyu Yue (Aluna), applied the NFT work of artist Jin Xiaoyao: Sunset at TV Tower (Size: 5500 pixels×9500 pixels, Date: 20230608) as the knitting pattern, and innovatively used the way of wrapping the dress form to present the effect of the garment, as well as the trailing part connected to the knitted pattern, forming a continuous image effect that echoes the work video. The production technology of the local knitting factory is used to empower the NFT work and realize its physical conversion from digital virtual to real, creating an unique visual experience. The knitting transformation was provided by Tongxiang Yongxin Clothing Co., Ltd.with material and technical support. After rounds of discussion and design, the color and graphic summarization of the work was carried out, then the professional knitting technology produced several test samples for the color of yarn, size specifications, process cycle, etc., to complete the transformation.

作品作者吕越采用艺术家金小尧NFT作品:《电视塔日落》(尺寸:5500像素×9500像素,创作日期:20230608),作为针织的图案,创新使用缠绕人台的方式呈现了服装效果以及拖尾部分与针织的画面连接,形成连续的画面效果,与视频作品呼应。作品用本地针织工厂的生产技术,赋能给NFT作品,并实现其数字虚拟到实物的物理转换,形成的整体视觉感受别有一番风味。

针织转化由桐乡市永欣服饰有限公司提供材质及技术支持,设计者经过多轮探讨与设计,对作品进行色彩及图形归纳,再由专业织造工艺针对纱线色彩、尺寸规格、工艺周期等进行了多次试验打样,完成转化。

Sunset Moment
Digital, knit

日落时分
数字、针织

He Ran/Zhejiang Huigang Fashion CO., Ltd. 赫然 / 浙江汇港时装有限公司

"Water benefits all things but does not compete". The Zen meaning of water is flowing in the spirit of traditional Chinese culture and philosophy. Water has become a representative of the integration between the universe and human beings, symbolizing the flow of life and the universe, and is endowed with a poetic meaning. The work uses knitting yarn to simulate the gurgling of a waterfall. The water is formless, but adapts to the shape of everything, stretching back and forth, but hiding power and opportunity; the water is silent, gentle, and soft, but tough and strong, accommodating everything. Layering, mingling, surrounding; surging, flowing, to one. Life is nurtured by water in a continuous stream that eventually gathers into millions of rivers and returns to nature for eternity. Just as the dark current surging with infinite vitality, and finally converge into "clothes".

"水善利万物而不争"，水的禅意，流淌在中国传统文化与哲学精神之中。水成为融合宇宙与人之间的代表，象征着生命和宇宙的流动，被赋予了诗意的含义。

作品以针织纱线模拟瀑布的潺潺水流，水无形，却适应万物之形，延绵回转，却暗藏力量与机遇；水无声，温润柔和却坚韧有力，容纳万物。层叠，交融，围绕；涌动，流淌，归一。生命在水的孕育中，源源不断地涌出，最终汇集成千万条河流，归向自然，成为永恒。正如暗流涌动，生机无限，终而汇聚成"衣"。

Surge
Mixed media (knitting yarn, beads, brushed stainless steel etc.)

涌动
综合材料（针织纱线、珠片、拉丝不锈钢等）

Bao Yiwen/Zhejiang Qianqiu Knitwear Co., Ltd.　　　　鲍怿文 / 浙江浅秋针织服饰有限公司

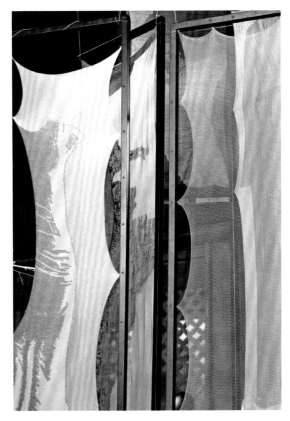

Screens have existed as a function of changing rooms for a long time, and the space they create represents transformation and renewal. Weakening the characters behind the screen is also a transformation of clothing language in the present. The stretching and deformation of clothing graphics are not equivalent to our body being controlled. The body behind the screen is no longer controlled, and this constraint will no longer come from the current clothing.

屏风作为更衣室的功用存在了很久,它形成的空间代表了转换和更新。把屏风后面的角色弱化也是服装语言在当下的一个转换。服饰图形的拉伸与变形并不等同于我们的身体为之受控。现在屏风后面的身体不再被控制,这种束缚也不会来自当下的服饰了。

Clothes as Screen
Mixed media

衣作屏
综合材料